ENGINEERED FABRICS

Edited by **Mukesh Kumar Singh**

Engineered Fabrics
http://dx.doi.org/10.5772/intechopen.74276
Edited by Mukesh Kumar Singh

Contributors

Han-Yong Jeon, Mario Riccio, Mauricio Ehrlich, Esin Sarıoğlu, Osman Babaarslan, Halil İbrahim Çelik, Münevver Ertek Avci, Dilan Canan Çelikel, David Branscomb, David Beale, Yang Shen, Royall Broughton, Vladimir Quiñones, Mukesh Kumar Singh

First published in London, United Kingdom, 2019 by IntechOpen
IntechOpen is the global imprint of INTECHOPEN LIMITED, registered in England and Wales, registration number: 11086078, The Shard, 25th floor, 32 London Bridge Street
London, SE19SG – United Kingdom
Printed in Croatia

British Library Cataloguing-in-Publication Data
A catalogue record for this book is available from the British Library

Additional hard copies can be obtained from orders@intechopen.com

Engineered Fabrics, Edited by Mukesh Kumar Singh
p. cm.
Print ISBN 978-1-78985-007-9
Online ISBN 978-1-78985-008-6

For EU product safety concerns:
IN TECH d.o.o., Prolaz Marije Krucifikse Kozulić 3, 51000 Rijeka, Croatia,
info@intechopen.com or visit our website at intechopen.com.

Meet the editor

Professor Mukesh Kumar Singh is a textile technocrat and academician with expertise in the expansion of novel methods for developing engineered fabrics by making alterations to weaving machines. His expertise also includes the development and study of a specific field of technical textile, i.e., cosmetotextiles. He has wide experience in engineering high-value polyester multifilament woven fabrics. Professor Singh obtained his PhD in Textile Engineering from the Department of Textile Technology, Indian Institute of Technology Delhi, New Delhi, India. He worked at Pasupati Spinning and Weaving Mills, Dharuhera Rewari Haryana, India, from 1992 to 1993, and then from 1995 to 1996 he worked first as a spinning assistant and later as a researcher at IIT Delhi under the research group of the late Professor Pushpa Bajaj and Professor Ashwani Kumar Agrawal from 1998 to 1999. Professor Singh joined the Government Central Textile Institute, Kanpur, as a lecturer in 1999. Professor Singh is included in the SCI Highly Cited in the Fields of Textile Technology and Textile Engineering with more than 40 papers. He wrote the book *Industrial Practices in Weaving Preparatory* by Woodhead Publications, 2014. He has contributed various book chapters on cosmetotextiles and photovoltaic textiles in books of international standards published by CRC Press and InTech Publications.

Contents

Preface

Engineered fabrics have attracted the attention of almost all facets of engineering due to their broad acceptability for applications in various fields. It is now the fastest growing sector of the textile industry and accounts for almost 19% (10 million tonnes) of the total world fiber consumption of all textile uses. This figure is likely to increase to 14 million tonnes by the year 2025.

The application of engineered fabrics increases day by day in civil engineering, automobile engineering, and healthcare sectors. The current volume of the worldwide market of technical textiles is more than $60 billion in which the share of engineered fabrics is approximately $30 billion. The average worldwide annual growth rate of technical textiles is expected to be around 4.2% for the period 2015–2020.

The uniqueness and challenge of engineered fabrics lie in the need to understand and apply the principles of textile science and technology to provide solutions for technological problems, but also engineering problems. With the emphasis on measurable textile performance in a particular field of application, this requires the technologist to have not only an intricate knowledge of fibers, yarns, and fabrics manufacturing, and textile science and technology, but also an understanding of the application of engineered fabrics. Thus, the consumer of acoustic textiles requires an intricate knowledge of sound and sound amplification, and the medical textile producer the requirements of a consultant, medical practitioner, and nurse. This book attempts to provide a bridge between producer and end-user, and will provide ample opportunity to the reader to understand the importance of engineered fabrics, their science, and technology.

Information in the book is collected from various sectors where engineered fabrics are utilized for technical textiles.

Each of the chapters has been specially prepared and dedicated to address a typical problem of engineered fabrics. This book covers so many developments registered in the field of engineered fabrics as well as future trends in the principles of manufacture and state-of-the-art constructional specifications, properties, test methods, and standards of the major product areas and applications of engineered fabrics.

A team of dedicated researchers and academicians has contributed a great deal of time, effort, and above all special and incredible expertise and experience to the preparation of this book. I wish to extend their sincere thanks to all the authors for their important contribution, patience, and cooperation to complete *Engineered Fabrics*.

In some chapters, typical sections are included with specific material on real case studies. These chapters will provide some research platforms to future researchers. This book includes up-to-date coverage on engineered fabrics. Each year exciting and valuable advances are registered in the field of engineered fabrics. While I understand that not all of these are appropriate for discussion in a specific book, I have incorporated the most up-to-date information and exciting, recent advances to maintain accurate descriptions of structures and processes and to illustrate essential points. Specific examples include a geotextile, textiles for acoustic applications, etc., and functions of such engineered fabrics.

The book will serve as a study guide for students that provides learning objectives, study outlines, and learning activities to help their science and engineering undergraduate and postgraduate course content.

This book once again confirms that enthusiasm and affection towards the subject of engineered fabrics are more important than any other gains. Special thanks are also given to IntechOpen for consistent effort and interest in keeping this project alive for a long duration and for having continued faith in the editors.

Professor Mukesh Kumar Singh
Director, UP Textile Technology Institute, Kanpur
Affiliated to Dr. APJ Abdul Kalam Technical University Lucknow, India

Chapter 1

Introductory Chapter: Engineered Fabrics

Mukesh Kumar Singh

Additional information is available at the end of the chapter

http://dx.doi.org/10.5772/intechopen.82717

1. Introduction

Engineered fabrics have become the need of present era because the application field of engineered fabrics have spread from automobile sector to aeronautics, marine to geo-engineering, sports items to packaging materials, etc., The present popularity of engineered fabrics is not an incident but it is a long journey which engineered fabrics have completed from triple layer fabrics to three-dimensional fabrics. Engineered fabrics also consists of solution-focused and custom-designed fabrics [1]. These products are also utilized in process industries outside of papermaking such as nonwovens, corrugators, building products, tannery and textile industries.

The growth of engineered fabrics is linked with application of both natural as well as man-made fibers. Engineered fabrics are becoming the base for various product developments for wide variety of applications [2]. Engineered fabrics are reaching to touch the 40–45% share of total fabric production in developed nations.

The supply chain of engineered fabrics follows a long route, starting from manufacturing and selection of appropriate fiber to manufacturing of specialty fabrics for engineering applications [3].

Although, the financial importance and justification of engineered fabrics spreads from conventional textile industry to almost all facets of human life still investors and manufacturers are not getting enough confidence to expand the production capacity at large scale. In spite of all these challenges, the field of engineered fabrics is very promising and only need to keep freshness in product development for better end uses.

Engineered fabrics cannot be developed by using only one type specialty fiber, yarn, weave and finish. This chapter belongs to consider various factors: commercial, technical and global which are major driving forces of this industry. Engineered fabrics have got attention from both side of Atlantic but China has registered remarkable growth in this sector and India is emerging at slow pace [4].

The engineered fabrics are used as raw material to serve various segments of technical textiles viz., agrotech, buildtech, cosmetotextiles, clothtech, hometech, indutech, mobiltech, sportech, packtech, meditech, protech, and others. The automobile textiles (mobiltech) segment is demanding highest amount of engineered fabrics followed by industrial textiles (Indutech). Various types of engineered fabrics like spacer fabrics, multilayer fabrics, needle punched nonwoven fabrics, melt blown nonwoven fabrics and warp knitted fabrics are highly demanded by various sectors of technical textiles [5].

The engineered fabrics are able to cater the needs of wide spectrum of present market starts from awnings, airbags, automobile filters, floor covering, fabrics used in erosion suppression, hoses, road construction, safety belts, thermal and sound insulation and upholstery, etc. Engineered fabric manufacturing industry is already established in strong position in China, India, Korea, Thailand and Taiwan. The engineered fabric market is continued to grow in coming years also. The growth of automobile, industrial sector and infrastructure sector are the major driving forces for engineered fabrics [6]. Being the world's second largest producer of textiles and apparel, India's engineered fabrics manufacturing sector is also growing at fast rate and creating both direct and indirect employment. The textile and garment industry is the root of Indian economy which provided employment to 105 million citizens. Indian textile industry will grow up to $223 billion by 2021 in which engineered fabric's sector will play major role. High transportation and energy cost and lack of labor reforms are some major hurdles in traditional Indian textile industry which force to shift its focus from conventional textile to engineered textiles. Export of engineered textiles is increasing with annual growth rate of 18%. Now, Government of India developed new policies for rapid growth of industry which will make remarkable change in engineered textiles. There are few steps taken to promote the engineered fabric manufacturing in India.

- Market development support to stabilize both domestic and international markets
- Investment promotion
- Exemption in custom duty for raw materials
- Implementation of uniform goods and service tax across the country
- Establishing standards for various types of engineered fabrics.

2. Definition

The Engineered fabrics are defined as "The fabrics which are produced by some modified fabric manufacturing techniques than conventional for unconventional engineering applications". Various critics and scientists will coin some other definitions in future also but the basic theme of engineered fabric may remain unchanged. Basically the engineered fabrics covers the 2D, 3D fabrics, belts, braided items, aerospace automotive textiles, industrial textiles, high performance textiles, etc. [7, 8].

The engineered fabrics can be comparable with composite materials also where two materials having different nature are combined together to extract the merits of both the materials in a single product, similarly two or more than two types of fibers, yarns, weaves or laying techniques are combined to engineer the targeted fabric [9]. In fact at this stage it is safe to say that any effort to define the engineered fabrics will prove insufficient because the development in this sector is in neonatal stage.

3. How does an engineered fabric differ from technical fabrics?

Since decades of years technical textiles was widely used to explain the unconventional textiles which includes bunch of fibers, ropes, cabled yarns, woven and nonwoven fabrics, finished fabrics, stitched textiles, etc. The term technical textiles is used to encompass all textile products other than those intended for apparel, household and furnishing end-uses, however, the term "engineered fabrics" is limited to various woven, nonwoven, knitted and braided fabrics manufactured by some unorthodox manufacturing techniques for special engineering applications. Various fabrics engineered for specific applications like medical, hygiene, sporting, transportation, construction, agricultural and many other purposes [10].

Engineered fabrics are used to provide the base for filters, machine clothing, conveyor belts, abrasive substrates, geofabrics, fabrics for acoustic and thermal insulation, etc. It is essential to mention that the composite materials made of polymeric membrane as reinforcing material with matrices, highly loose structured materials such as chopped strand mat, milled glass and pulped organic fibers cannot become the part of engineering fabrics [11].

4. Suitable raw material for engineered fabrics

Various natural fibers have enough potential to become the part of engineered fabrics. The major natural fibers have been used as basic material in engineered fabrics is cotton, flax, jute and sisal. These fiber are used to manufacture various heavy engineered fabrics like canvas, needle punched nonwoven fabrics for geo applications, ropes, belts and other multilayer fabrics, etc. [12]. However, some limitations of these fibers restricted the growth in engineered fabrics in which higher rigidity, prone to fungal and microbial attack; poor water resistance and lower flame retardancy are remarkable. Jute is cheaply available fiber which has ample potential to be used in engineered fabrics in gray and treated form. Sisal fiber is suitable material for ropes, nets and twines manufacturing [13].

Wool is another natural option with merits of higher limiting oxygen index value, thermal insulation but its limited availability and versatility has restricted its applications in engineered fabrics [14]. Silk fiber is another rare option for engineered fabrics due to its low availability and higher cost [15].

4.1. Regenerated fibers

First commercially manufactured manmade fiber developed 1905–1910, is still suitable material for manufacturing engineered fabrics like tyre cords, preforms for conveyer belts and hoses, etc. Some other regenerated fibers like acetate rayon and cuprammonium rayon also have found its place in engineered fabrics [16].

4.2. Synthetic fibers

4.2.1. Polyolefins

Polyethylene (PE) and polypropylene (PP) are two major fibers of this group which have registered its valuable presence in the manufacturing of engineered fabrics. Low density, easy manufacturing techniques, high moisture and abrasion resistance have secured its rapid growth in engineered fabrics. The major engineered fabrics made of these fibers are used to manufacture bags, carpet bases, furniture linings, sacks, nets and other marine textiles. PP Fiber has good wicking with poor moisture absorption potential and this characteristic make this fiber appropriate for use in engineering of high performance diapers. The PP fiber has low spinning temperature (210–220°C) have proved ideally suited material for meltblowing and spun bonding techniques to manufacture engineered nonwoven structures quickly [17].

4.2.2. Polyamide

Polyamide fiber group containing various nylon fibers like nylon 6, nylon 66, nylon 6.10, etc. have good abrasion resistance, high strength, remarkable elasticity and excellent impact absorbing potential proved very useful in manufacturing various engineered items like parachute fabrics, spinnaker sails, reinforced tyres and geofabrics for high performance road construction. Western Europe and North America are more strongly inclined towards nylon 66 while Asia and Eastern Europe produce predominantly nylon 6 [18].

4.2.3. Polyester

Polyester is low cost fiber with plenty of merits like high abrasion resistance, high strength, low moisture regain and excellent uniformity. Recycled polyester fiber is another cost effective alternative fiber for manufacturing of engineered fabrics like spun bonded structures, needle punched structures, etc. [19]. A modified polyester fiber is used widely in manufacturing of flame retardant fabrics, waterproof breathable fabrics and canvas fabrics.

4.2.4. Glass and ceramics

Glass fiber was very difficult handle for many years, been one of the most underutilized fibers. This fiber is used in various engineered nonwoven structures to be considered as a cheap insulating material and reinforcement preforms for relatively low performance composites like fiber glass and heat-resistant materials. The applications of glass fiber increasing day by day in the form of engineered structures for sealing materials, rubber reinforcement, as well as filtration, protective clothing, packaging metal body parts and components [20]. Some ceramic fibers have found limited applications in engineered structures due its high cost and poor bending performance.

4.2.5. High performance fibers

4.2.5.1. Poly(amide-imide) fibers

Successful polyamide-imide fiber was produced by Rhone-Poulenc Inc. with a trade name of Kermel. The limiting oxygen index (LOI) of Kermel fiber is 32. It remains safe without any degradation up to 250°C for a exposure of 500 h to heat. This fiber does not have melting temperature Tm but is carbonize. Kermel fiber can be blend successfully with other commercial fibers like viscose and polyester. A wide variety of engineered fabrics with Kermel fiber can be produced for air forces, army, navy and firefighter dresses [21].

4.2.5.2. Polybenzimidazole (PBI) fibers

The PBI fiber was invented by Celanese Inc. This fiber is highly stable at 300–350°C. Its limiting oxygen index (LOI) value is 41, which is quite safe and higher than threshold value 25. This fiber offer equal heat protection to asbestos with half density. It has moisture regain. The PBI fiber based engineered fabrics are used as reinforcing material to produce fire protection in aircraft seats, firefighter suits and racing-car driver suits. It found its smart applications in in rocket motors and boosters to provide safety against ignition [22]. The engineered fabrics made of PBI fibers offer excellent resistant to puncturing, tearing and ripping.

4.2.5.3. Phenolic or novoloid fibers

Phenolic or novoloid fibers fiber is manufactured by spinning and postcuring of phenol formaldehyde resin precondensate. Kynol is a well-established novoloid heat-resistant fiber of GUN EI chemical industry. Kynol fiber is golden in color, soft feel with moisture regain of 6%. It slowly carbonized at very high temperature without any smoke. It has poor strength and abrasion resistance which suppresses it application in apparel sector. It can be easily blended with aramid fibers like nomex to make it suitable for flame retardant apparel applications. Philene is another important fiber member of this group with moisture regain of 7.3% and LOI 39% [23].

4.2.5.4. Modacrylic

The modacrylic fiber still has first choice of manufacturers to engineer flame-retardant fabrics. Modacrylic fibers are produced under various commercial names, such as SEF (Solutia Inc.), Velicren FR (Montefibre, Italy), Elura (Monsanto Fibers), Dynel (Union Carbide) and Verel (Tennessee Eastman). Modacrylic fiber and is a copolymer of acrylonitrile, vinyl chloride or vinylidene chloride in the ratio of 60:40 (w/w) along with a sulfonated vinyl monomer. Modacrylic fiber has LOI in the range of 26–31%. Kaneka Corporation has also developed Kanecaron, an FR modacrylic with an LOI value in the range of 30–35%. Fabrics from Kanecaron with commercial name of Protex M has LOI 33% blended with cotton, while maintaining the softness and comfort similar to cotton fabric.

5. Engineered fabrics

Engineered fabrics are textile materials manufactured primarily for technical and functional performances. Most of the engineered fabrics are manufactured by assembly of fibers, yarns and/or strips of material which have a very high surface area in comparison to their thickness and have sufficient mechanical strength. Engineered fabrics are commonly manufactured by weaving, knitting, felting, lace making, nonwoven processes, net making and tufting or a combination of these processes. Most of the engineered fabrics are two dimensional structures but recently three-dimensional structures have become very popular structure in this segment. The knitted structure consist one set of thread, woven consist two set of threads in the form of warp and weft but three-dimensional structure consist three set of threads: warp, weft and stuffer thread.

5.1. Weave structures

The two dimension engineered fabrics consists various weaves in which plain and leno weaves are widely used. There are some others weaves which can be proved functionality in engineered fabrics. All threads do not follow the straight path in woven structures and consist a crimp [24].

5.1.1. Plain weave and derivatives

The simplest weave to manufacture engineered fabrics is plain weave which is produced by alternatively lifting and lowering one warp thread across one weft thread. The performance of engineered fabrics has plain weave will depend type of fiber used: either staple or filament, type of yarn: flat, textured and twisted, yarn linear density and fabric set. The bending rigidity of engineered fabrics depends on the stiffness of the raw materials used and by the twist factor of the yarn and thread density in woven fabric [25]. Amount of twist in constituent yarns of engineered fabrics is used to impart specific features like extensibility, surface roughness and texture, etc. By changing the areal density (fabric grams per square meter, GSM) and cover factor affect the abrasion resistance, dimensional stability, filtration potential, porosity, stiffness, strength and thickness of engineered fabrics can be altered [26]. Square sett plain woven fabrics that are fabrics have nearly the constant number of ends and picks per unit space and warp and weft yarns of the same linear densities are produced with similar cover factors. Light weight plain woven fabrics with lower areal density and low cover factor with open weave construction are used as bandages and cheese cloths while highly open cloths are used in geotextile stabilization fabrics and heavy closely woven fabrics include cotton awnings.

Plain weave can be modified in the form of Rib and Matt weave. These weaves are produced when two or more than two adjoining warp or weft threads are considered as one unit and lifts or downs simultaneously. These weaves gives a higher cover factor, without jamming the weave structure [27].

Simple matt (or hopsack) woven fabrics offer a similar texture to plain woven fabric. The simplest matt weave is a 2/2 matt where two warp ends are lifted over two picks (unit of two warps and two weft act as a unit in plain weave). The unit of lifting threads can be increased to 3 or 4 to create 3/3 or 4/4 matt weave structures.

Some typical matt weaves, like a 4/2 matt, are produced to obtain special engineered effects.

Plain weave can be modified in another way in which either the ends or picks keeps more with higher crimp is called rib structure. If the number of ends is more than picks per unit length with high warp crimp, it is called as warp rib and vice versa for weft rib fabrics [28].

5.1.2. Triaxial weaves

Almost all two-dimensional woven structures have been developed from plain weave fabrics in which warp and weft yarns are interlaced at 90° or at nearly 90°. The triaxial fabrics are the only exception, where two sets of warp yarns are generally inserted at 60° to the weft. In case of tetra-axial fabrics, four sets of yarns are inserted at 45° to each other. Triaxial fabrics are manufacturing on commercial machines. The first triaxial weaving machines were developed by the Barber Colman Co. and further developed by Howa Machinery Ltd., Japan. Triaxial fabrics can be defined as set of threads where the three sets of threads form a multitude of equilateral triangles in which two sets of warp yarn are interlaced at 60° with each other and with the weft. The tearing and bursting strength of triaxial fabrics is remarkable higher than conventional fabrics. The shear rigidity of triaxial fabrics remains superior due to locked intersection points. Triaxial woven engineered fabrics have found a wide range of technical applications in, balloon fabrics, pressure receptacles, sailcloths, tyre fabrics and laminated structures [29].

5.1.3. Three-dimensional woven engineered fabrics

Three-dimensional woven engineered fabrics are produced to enhance the strength, thickness, extensibility, porosity and durability in woven engineered fabrics.

The performance of 3D woven fabrics can be engineered by making some alteration in weave used, the thread spacing, raw materials structure (filament or staple), linear density (or count) and twist factors of the warp and weft yarns. There are countless possibilities in 3D woven engineered fabrics to manufacture engineered fabrics of desired properties [5].

Engineered fabrics manufacturing processes: the essential operations in the weaving of a cloth are:

- Shedding, i.e. the separation of the warp threads into two (or more) sheets according to a pattern to allow for weft insertion
- Weft insertion (picking)
- Beating-up, i.e. forcing the pick, which has been inserted into the shed, up to the fell of the cloth (line where the cloth terminates after the previous pick has been inserted).

Secondary motions are incorporated to make the provision for the supply of warp and weft warp yarns and for the cloth. The warp yarn is usually supplied from warp beam(s) and the weft yarn from the pirn on shuttle looms only or cones on shuttles looms. Most of the single phase weaving machines uses same kind of motions and an almost horizontal warp sheet between the back rest and the front rest. Such kind of system is utilized in common shuttle looms, rapier looms, projectile looms, air jet looms and water jet looms [30].

6. Engineered fabrics by nonwoven fabric manufacturing

It is difficult to define the nonwoven fabrics because country wise definitions of nonwoven are available which have very poor coherence with other. However the most acceptable definition was coined by the American Society for Testing Materials (ASTM D 1117-80). Although this definition solved the limited purposes to define the nonwoven. The nonwoven fabrics can be redefine as "A nonwoven textile structure can be produced by bonding, interlocking, intermingling, pressing of textile fibers or in combination by means of mechanical, chemical or thermal techniques and their combinations by shortening of conventional fabric manufacturing processes" . The nonwoven fabric manufacturing can be divided into two sections. The first section is dedicated for fiber web manufacturing and second section for bonding or interlocking of constituent fibers, the layering of various webs one over another in various fashions which decides the nonwoven structure properties up to major extent is called batt. The batt is subjected to bonding or interlocking process for final product manufacturing [31].

7. Batt production by carding machines

The main objective of carding process is individualization of fibers after removing short fibers up to some extent but the carding machines for nonwoven batt production have some modifications like two cylinders in place of one in conventional cards. In case of nonwoven engineered fabric production carding process is nearly final process because after carding the chances of fiber blending goes to zero. Generally short-staple revolving flat cards are most suitable for nonwoven industry due to its high opening potential with high production rate. These cards are equipped with autoleveller facility to improve the uniformity in mass per unit length of web. The card web has very low web density and high degree of variation in mass per unit length which is not suitable to be used directly in a nonwoven. There are three main way to lay the web during batt formation: parallel laying, cross laying and bias laying [32].

7.1. Parallel laying

The parallel laying is the basic, cheapest and simplest way of batt formation. In this system numbers of cards are situated one above another or side by side slightly above the main conveyor belt. The webs from each card came down onto the batt forming conveyor lattice with number of times (number equals to the card numbers) the mass per unit area. The card webs are turned through a right angle with the help of a guide which turns the web at 45°. These techniques provide maximum number of fiber lying along the batt direction which is called machine direction and very few remains across the batt direction. This type of web can be converted to engineered nonwoven fabric by opting anyone way of either bonding or entanglement. The strength of bond in parallel laid nonwoven remains less than individual fiber strength. The parallel laying process suits to manufacture narrow tapes and medical textiles while cross laying suits to filter and wipe fabrics. However randomized doffer cards neutralize the situation up to major extent by distributing the fibers randomly together with 'scrambling rollers'. Both parallel laid and cross laid laying shows anisotropic behavior, however by combining both parallel laying and cross laying isotropic nonwoven structures are engineered.

The final width of nonwoven engineered structure is a challenge and it can be overcome by combining various laying techniques [32].

7.2. Cross laying

In order to result cross laying of webs to form batt, the cards are kept at right angles to the main conveyor lattice M and the card web is moved backwards and forwards across the main moving conveyor lattice.

The speed of main conveyor lattice is kept slow to accommodate many layers of card web in desired order. The cross laying systems suffers with two major problems; first, this system prone to form heavier batt at the edge due to overlapping. This issue can be solved by moving the of direction of batt at the edge of lattice. The second is to match the input speed of cross laying with card web speed. Generally input speed remains less and card web speed must reduce to match with input speed.

7.3. Wet laying

This technique of batt formation is influenced by paper making industry. The fibers are dispersed into water and water content is kept sufficient to prevent fiber aggregation. This system promotes the blending of fibers and laying them successfully. Wood pulps can also be blended with fibers to form the batt. This system is suitable to the batt of wooden pulp and fibers used in sanitary napkin manufacturing. The wet-laid batt is used in some other disposable engineered products like drapes, gowns, sometimes as sheets, as one-use filters, and as coverstock in disposable nappies [33].

7.4. Spun laying

This technique of batt formation offers shortest route. This includes extrusion of the filaments from extruder, drawing the filaments and laying them in the form of batt. At the same time bonding also takes place which makes this process very economic from polymer to fabric manufacturing cost point of view. Initially, this process was developed for large scale production but at present small size machines are available to cater the need of small scale manufacturers. Initially polyester and polypropylene fibers were spun-laid but presently polyamide and polyethylene fibers can also be processed on this system. The microfiber technology also integrated with this system which enhanced the versatility to produce finer, softer and better filtration engineered fabric structures. The process starts from feeding of polymer chips into extruder which feeds the molten mass of polymer to a metering pump and then to a group of spinnerets which quenched further for quick solidification. The drawing process is assisted by hot air blowing in this system. The fiber orientation is controlled by both the direction of filament delivery tube and conveyor belt to assure uniform distribution of fibers [34].

7.5. Air laying

The air-laying system is capable to offer the desired batt in single stroke at high speed without first making lighter weight web and then by laying. The fiber opening potential of this system is limited and needs ample pre-opening before feed to air laying system. This system consist opening and blending section in back of feed hopper which is used to deliver fiber sheet to the feed rollers. The fibers are then taking-off by consist fine wire metallic clothing on its surface, revolves at high speed. Some optional stripping rollers may attach to enhance the opening potential of the system. The opened fibers are removed by powerful air stream from opening cylinder surface. The air stream carries the fibers to cage like conveyor lattice to form the final batt [35].

7.6. Melt blown

The melt blowing process is another very promising method of manufacturing very fine deniers. This system produces fibers without the use of fine orifice spinnerets at high production rate. In this arrangement polymer is melted and extruded normally as other melt extrusion processes but through relatively large spinneret orifices. After complete melting, filtration, polymer melt extrude out from spinneret orifices it directly comes in the contact of very high temperature (above the melting temperature of polymer, Tm) hot air stream which assist in filament stretching up to maximum extent. The staple fibers of very fine deniers produced in this way are collected on the surface of permeable conveyor to form a batt as in air laying and spun laying.

Bonding is rarely required here and in most of the cases the melt-blown batt is laminated on another nonwoven structure (may be a spun-laid or the melt-blown batt). This type of laminated engineered fabric is used to engineer breathable protective clothing for use in agriculture hospitals and industry. These structures are useful as battery separators, industrial wipes and clothing interlinings with good insulation properties also. If melt blown layered structure is not bonded and directly collected as nonwoven batt then it is used as ultrafine filters for air conditioning and personal face masks, oil-spill absorbents and personal hygiene products. This technique is growing with 10% annual growth rate [36].

7.7. Chemical bonding

Chemical bonding is the process of sticking fibers of batt by treating/modifying either a specific area of batt or whole batt. A variety of bonding agents/adhesives are available in which acrylic latex, styrene butadiene lattices and vinyl acetate latex are the major one. The bonding agent must have ample wettability otherwise it can be maintained by adding appropriate amount of surfactants [37]. After judicious application of bonding agent, the batt is dried then to remove aqueous component and making proper bonding among the fibers of that localized region. Finally, the treated batt is cured at higher temperature to develop crosslinks both inside and between the polymer particles at 120–140°C for 2–4 min.

7.8. Thermal bonding

This technique of bonding is tagged as eco-friendly because the application of any kind of chemical is negligible. Productivity of thermal bonding process remains higher than any other chemical bonding process. Thermal bonding process is energy efficient also because it saves the energy which consumes to evaporate water from the binder and curing. Thermal bonding strategy can be divided into three classes like in first all of the fibers of same type with common melting behavior, second; a blend of fusible (lower melting point) and non-fusible (either the higher melting point or non-melting fibers) fibers and third; by application of bi-component fiber in which one component is fusible and other component is non-fusible. The temperature is applied at a localized area with or without pressure to melt the fusible fiber component and to stick with non-fusible fibers [38].

7.9. Spray bonding

In this technique latex binder is sprayed which act as bonding element to bind the fibers. There may be more number of spray cycles depending upon desired bonding extent and batt thickness because every spray cycle reduces the batt thickness up to some extent. These engineered fabrics can be used as raw material for hometech sector as quilt filling material, duvets and some typical type of filters [39].

7.10. Foam bonding

In order to reduce the application of water in various bonding techniques which not only enhances the cost of manufacturing due to essential drying but also the risk of binder migration, the foam bonding is better alternate in this direction. A definite amount of compressed air is passed through binder solution to create foam and then it applied on both side of batt with the help of horizontal nip of the impregnating roller. Foam consist limited amount of binder and negligible water content which suits for targeted application for bonding point of view.

7.11. Print bonding

This technique is used to apply the binder on one or both side of batt to limited portion and in a set pattern. In order to assure penetration of binder well inside the batt, it is first impregnated with water and then binder is printed on batt in defined pattern either a printing roller or a rotary screen printer. The ratio of printed/unprinted area decides the ultimate properties of final nonwoven engineered fabric. The limited application of binder in print bonded fabric keeps fabric soft and pleasant feel. Print pattern and print content decides on the basis of type of fiber, fiber orientation and other properties of fibers used in the batt. Print-bonded fabrics have found its application in disposable/protective clothing, coverstock and wiping cloths.

7.12. Powder bonding

Powder bonding technique is based on the application of thermoplastic powders alternate to thermoplastic fibers. Rest processes remain similar to thermobonding. The powder bonded engineered fabrics show better flexibility and softness with poor bonding strength. These structures are used in protective apparel and coverstock areas where high bulk is desired.

8. Engineered fabrics by fiber entanglements

There are three methods of producing engineered fabric by fiber entanglements; needle punch, hydroentanglement and stitch bonding. These three methods are based on fiber entanglements and frictional behavior of fibers and conceptually known as mechanical bonding. Out of these three techniques needle punch is most popular and simplest one [40].

8.1. Needle punching

The concept of needle punching is quite clear and simple. In this method the batt is passes between two stationary plates, the bed and stripper plates. While between the plates the batt is penetrated. The needle density remains up to about 4000 m^{-1} width of the loom. The design of penetrating needle plays major role in fiber entanglement. Needles are generally made triangular in shape and have barbs cut into the three. As the needle goes down into the batt the barbs traps some fibers and pull them through the other fibers to get it entangled.

When the needles return back in upward direction, the fiber loops formed during downward movement of needles tend to remain in position, because they are released by the barbs. This downward penetration of needles takes place repeatedly which makes the batt much denser and finally needle punched structure manufactured [41].

8.2. Hydroentanglement

The hydroentanglement process of engineered fabrics manufacturing was developed by DuPont in 1960. This process is quite similar to needle punch process. This technique is used to entangle the fibers of lightweight batt. In this process very fine nozzles are used to inject the water in the form of fine water streams or droplets. Number of fine nozzles is situated at the edges of batt. The water stream passes through the perforated screen to remove the used water. The fiber which come in the contact of water get wetted and its total momentum goes compare to other fibers and these fibers get entangles with other fibers of the batt. Water cleanliness, pH and temperature are critical issues to be taken care during the manufacturing. This process is capable to produce engineered fabrics for wipes, surgical gowns, disposable protective clothing and backing fabrics for coating applications [42].

9. Engineered fabric manufacturing by weaving

Weaving is most popular promising technique of engineered fabric manufacturing. Presently shuttle looms are obsolete and out of the international manufacturing scene.

9.1. Rapier looms for engineered fabrics manufacturing

Rapier was the first concept that successfully replaced the shuttle weft insertion system. First generation of Rapier looms did not get commercial acceptance due to its very low speed. With the invention and introduction of precision engineering and microprocessor controls, the weft insertion rates have increased remarkably.

The Rapier loom of 2.5 m width has close competition with projectile loom. The single rapier looms are rigid rapier slow speed looms. However, the invention of double rapier has increased the commercial acceptability because wide variety of threads can be processed on these looms. Both rapier enter from both extreme end of reed and meet at the middle of cloth width to transfer the weft thread from one rapier to other rapier. Rapier looms have two weft insertion systems; one is Gabler and other is Dewas system. In case of Gabler weft insertion system weft is inserted alternately from both sides of the machine [43].

The weft thread is cut every second pick with hairpin selvedges being formed alternately on both selvedges but weft is inserted from one end of rapier loom in Dewas system. Dewas system is dominating now a days and most of the looms has weft feeding system on one side. Double rapier weaving machines may have either the rigid or flexible rapiers. Dornier HTV and P19 series Rapier looms are capable of weaving most of the industrial fabrics with weft linear densities of up to 3000 tex, in loom widths of up to 4600 mm and at weft insertion rates of up to 1000 m min^{-1}. Rapier looms are used widely to manufacture wide range of engineered fabrics starts from opencoated geotextile mesh, heavy conveyor belt cloths, home textiles, and canvas and furnishing items. Rapier looms are most suitable weaving machines to carry and run Jacquard shedding device.

9.2. Projectile looms

The first projectile weaving machine was based on single projectile which had provision to strike the projectile from each side of the loom. This machine had weft supply system from both side of the loom. The latest projectile looms have multiple projectiles which are stroked from one side and are returned back to the picking position with the help of a conveyor belt. The contribution of Sulzer Textile to develop projectile loom and enhanced its versatility in terms of improved weft insertion rates, machine efficiency and extended the range of fabrics manufactured is unforgettable. Projectile loom offers facility to use a winding cone directly without rewinding which saves cost and time both. The length of standard projectile is 90 mm with 40 g weight. The weft thread is withdrawn from weft supply cone through a weft brake and a weft tensioning device to the weft feeder which places it into the gripper of the projectile [44].

A torsion rod system is used for picking which transfers the maximum possible strain-energy to the projectile before it leaves the picker shoe. The strain energy can be adjusted by changing the position of torsion bar. Sulzer Textil redesigned the reed of projectile loom which offer more effective and strong beat-up. A weft insertion speed of 1300 m min^{-1} can be achieved on 3600 mm reed width machine. Latest projectile looms are capable to insert six color weft threads, fancy threads and wide variety of material from fine polyester to coarse woolen threads successfully. The machines can be equipped with a variety of shedding mechanism like dobby and jacquard. Machine performance can be monitored with microprocessors. Sulzer Ruti and Jäger are two major manufactures of projectile loom. Jäger have developed a hydraulically propelled projectile loom. Projectile looms are capable to weave wide variety of engineered fabrics of up to 8 m width, for awnings, airbags, conveyor belts, geotextiles, sailcloth, tyre cord fabrics, and a wide variety of filter fabrics of varying area density and air permeability.

9.3. Air-jet loom

The major aim of product development in woven fabric is to engineer new fabric structures having the most appropriate properties to achieve a high level of performance with suitable quality. In air jet loom weft thread is accelerated and passes through the shed by the flow impedance between the flowing compressed air and the weft. The energy creating from compressed air supplied from the compressed air tank to the air-nozzles reserves the kinetic energy in the nozzle, which accelerates and passes the weft through the shed. The compressed air leaving the nozzle combines with atmospheric air, it disperses, and the axial speed of compressed air drops quickly as it moves away from the nozzle. Therefore, in order to achieve wider loom width on air-jet loom, the compressed air speed must be maintained up to carry the weft thread. Three different systems have been adopted by commercial air jet loom manufacturers: single nozzle with confusor guides, multiple nozzles with guides and multiple (relay) nozzles with tunnel reed. Multiphase weaving machines have also adopted air-jet weaving concept. At present, the air-jet looms are very versatile and capable to process wide variety of weft threads. Hence, it become most suitable machine for engineered fabric manufacturing with weft insertion speed of 1000–2500 m min^{-1} [45].

10. Challenges and barriers

Designing and promotion of engineered fabrics is remarkable challenge in this sector. The conclusions can be arranged under following points:

- Protectionist policies of some countries are creating big hurdles in free flow of investment, technology and engineered fabrics products
- Lack of automation and dependency on conventional fabric manufacturing machineries
- Lack of skilled worker
- Lack of promotion of engineered fabrics
- There are enough potential of growth in engineered fabrics because the areas of applications are countless
- Engineered fabrics have found its place from inside the earth, deep under sea to high in the sky.

Author details

Mukesh Kumar Singh[1,2]

Address all correspondence to: mukesh70ster@gmail.com

1 Uttar Pradesh Textile Technology Institute, Kanpur, India

2 Dr APJ Abdul Kalam Technical University, Lucknow, India

References

[1] Stoppa M, Chiolerio A. Wearable electronics and smart textiles: A critical review. Sensors. 2014;**14**:11957-11992

[2] Singha K. A review on coating & lamination in textiles: Processes and applications. American Journal of Polymer Science. 2012;**2**(3):39-49

[3] Gupta D. Functional clothing—Definition and classification. Indian Journal of Fibre & Textile Research. 2011;**36**(4):321-326

[4] Kniivilä M. Industrial Development and Economic Growth: Implications for Poverty Reduction and Income Inequality. Helsinki, Finland: Pellervo Economic Research Institute; 2016. pp. 295-320

[5] Bilisik K, Karaduman SN, Bilisik NE. Chapter 4, non-woven fabrics. In: 3D Fabric for Technical Textile Applications. London, United Kingdom: Intech Publications; 2016. pp. 82-100. DOI: 10.5772/61224

[6] Singh J. Analysis of Trends in the Manufacturing Growth in Last Five Years. New Delhi, India: Research Studies Office of the Economic Adviser Department of Industrial Policy and Promotion Ministry of Commerce & Industry Udyog Bhawan; 2013. pp. 4-58

[7] Castano LM, Flatau AB. Smart fabric sensors and e-textile technologies: A review. Smart Materials and Structures IOP Publications. 2014;**23**:1-27

[8] Axisa F, Schmitt PM, Gehin C, Delhomme G, McAdams E, Dittmar A. Flexible technologies and smart clothing for citizen medicine, home healthcare, and disease prevention. IEEE Transactions on Information Technology in Biomedicine. 2005;**9**:325-9336

[9] Hewitt JA, Brown D, Clarke RB. Computer modeling of woven composite materials. Composites. 1995;**26**:134-26140

[10] Shishoo RL. Technical textiles; technological and market developments and trends. Indian Journal of Fibre & Textile Research. 1997;**22**(4):213-221

[11] Composite Materials Handbook, Department of Defence Handbook. Emerging Markets—Drivers and Hindrances of Sustainable Economic Growth [University of Agder masters thesis]. 2016. pp. 10-100

[12] Bavan S, Channabasappa MKG. Potential use of natural fiber composite materials in India. Journal of Reinforced Plastics and Composites. 2010;**29**(24):3600-3613

[13] Azam A, Shaker K, Nawab Y, Jabbar M, Hussain T, Militky J, et al. Hydrophobic treatment of natural fibers and their composites—A review. Journal of Industrial Textiles. 2016;**47**(8):1-44

[14] Frydrych I, Dziworska G, Bilska J. Comparative analysis of the thermal insulation properties of fabrics made of natural and man-made cellulose fibres. Fibres and Textiles in Eastern Europe. 2002;**10**:40-44

[15] Wakelyn P. Cotton Fiber Chemistry and Technology. Boca Raton: CRC Press; 2007

[16] Lipp-S B, Sztajnowski S, Wojciechowsk D. New commercial fibres called 'bamboo fibres' their structure and properties. Fibres & Textiles in Eastern Europe. 2011;**19**(1):18-23

[17] Bo Z. Production of PP meltblown fabrics-II. Indian Journal of Fibre& Textile Research. 2012;**27**(4):326-330

[18] Mukhopadhyay SK. Manufacturing, properties and tensile failure of nylon fibres. In: Bunsell AR, editor. Handbook of Tensile Properties of Textile and Technical Fibres. Cambridge, UK: Elsevier Publication; 2009. ISBN 978-1-84569-387-9

[19] Kellie G. 7-Developments in composite nonwovens and related materials. Advances in Technical Nonwovens. Doxford, UK: Woodhead Publishing; 2016: 213-26

[20] Landesmanna A, Alexandre SC, Batistaa E de M. Mechanical properties of glass fiber reinforced polymers members for structural applications. Materials Research. 2015;**18**(6):1372-1383

[21] Bourbigot S, Flambard X. Heat resistance and flammability of high performance fibres: A review. Fire & Materials. 2002;**26**:155-168

[22] Perri V. M.Tech thesis. Melbourne, Australia: RMIT University; 2016. pp. 5-50

[23] Bajaj P. Fire retardant materials. Bulletin of Materials Science. 1992;**15**(1):67-76

[24] Pan N. Analysis of woven fabric strength: Prediction of fabric strength under uniaxial and biaxial extensions. Composites Science and Technology. 1996;**56**:311-327

[25] Dhoot NS, Patil LG, Katkar PM. Effect of fabric weaves on compressional behavior of woven fabric. Indian Journal of Fibre & Textile Research. 2014;**39**(1):79-82

[26] Kaynak HK, Topalbekiroğlu. Influence of fabric pattern on the abrasion resistance property of woven fabrics. Fibres & Textiles in Eastern Europe. 2008;**16**(66):54-56

[27] Sekerden F. Effect of fabric weave and weft types on the characteristics of bamboo/cotton woven fabrics. Fibres & Textiles in Eastern Europe. 2011;**19**(6):47-52

[28] Mathur K, Seyam A-Fattah M. Color and Weave Relationship in Woven Fabrics. Raleigh, North Carolina: NCSU University Pub; 2011. pp. 129-150

[29] Rohrs JD. Design of a Bias-Weaving Machine: Thread Manipulation and Other Topics [M.Tech thesis]. Cambridge, USA: MIT; 2003. pp. 3-53

[30] Kumar B, Hu J. Woven fabric structures and properties. Engineering of High-Performance Textiles. 2018;**6**:175-230

[31] Behera BK, Militky J, Mishra MR, Kremenakova D. Modeling of woven fabrics geometry and properties. In: Jeon H-Y, editor. Woven Fabrics. London, UK: InTech Open Publication; 2012. ISBN: 978-953-51-0607-4

[32] Sawhney P, Reynolds M, Condon B, Slopek R, Allen C. A comparative study of nonwoven fabrics made with two distinctly different forms of greige cotton lint. Textile Research Journal. 2011;**00**(0):1-9

[33] Kalebek NA, Babaarslan O. Chapter 1, nonwoven fabrics. Fiber Selection for the Production of Nonwovens. London, United Kingdom: InTech Open Pub; 2016. pp. 1-28

[34] Lin H. A review of spunbond process. Journal of Textile and Apparel, Technology and Management. 2010;**6**(3):1-13

[35] Hubble MA, Koukoulas AA. Wetlaid nonwoven manufacture. BioResources. 2016;**11**(2): 5500-5552

[36] Duran D. Investigation of the physical characteristics of polypropylene meltblown nonwovens under varying production parameters. In: El-Sonbati A, editor. Thermoplastic Elastomers. London, UK: InTech Open Publication; 2012. pp. 243-260. ISBN 978-953-51-0346-2

[37] Kritzer P, Cook JA. Nonwovens as Separators for Alkaline Batteries. Journal of Electrochemical Society. 2007;**154**(4):A481-A494

[38] Desai AN, Balasubramaniam N. Critical factors affecting the properties of thermal-bonded nonwovens with special reference to cellulosic fibres. Indian Journal of Fibre & Textile Research. 1994;**19**(3):209-215

[39] Chinta SK, Landage SM, Krati Y. Application of chicken feathers in technical textiles. International Journal of Innovative Research in Science, Engineering and Technology. 2013;**2**(4):1158-1165

[40] Zhang Y, Deng C, Qu B, Zhan Q, Jin X. 5th Asia Conference on Mechanical and Materials Engineering. Pilsen, Czech Republic: IOP Publishing; 2017. p. 241

[41] Midha VK, Mukhopadhyay A. Bulk and physical properties of needle punched nonwoven fabric. Indian Journal of Fibre & Textile Research. 2005;**30**(2):218-229

[42] Mbwana SN, X-yu J, Chen T, Yu C. Optimum water jets inclination angle for better tensile strength in hydroentanglement process. Fibres & Textiles in Eastern Europe. 2009;**17**(4):82-86

[43] Maity S, Singha K, Singha M. Recent developments in rapier weaving machines in textiles. American Journal of System Science. 2012;**1**(1):7-16

[44] Talukdar MK, Arora A. Effect of weft insertion system on physical properties of fabrics. Indian Journal of Fibre & Textile Research. 1992;**17**(2):65-68

[45] Turhan Y, Eren R. The effect of loom settings on weavability limits on air-jet weaving machines. Textile Research Journal. 2018;**82**(2):172-182

Chapter 2

Denim Fabrics Woven with Dual Core-Spun Yarns

Osman Babaarslan, Esin Sarioğlu,
Halil İbrahim Çelik and Münevver Artek Avci

Additional information is available at the end of the chapter

http://dx.doi.org/10.5772/intechopen.80286

Abstract

Elastic core-spun yarns which is used as weft yarn for textile fabrics gained great importance in the last decade its due to the fact that stretch and recovery, comfort fits and flexibility properties. The technological progress made the dual core-spun yarn production possible. The dual core-spun yarns are composed of filament that contributes durability and polyurethane based elastane that provides stretchability to the fabrics. Hereby, both filament and elastane characteristics have great influence on denim performance at the same time. The main purpose of this study is to achieve the effect of filament fineness and elastane draft on denim fabric performance such as breaking force, breaking elongation, tear force, vertical elastic recovery, moisture management that is wicking rate and water absorption properties. Meanwhile, filament core-spun yarns with different filament fineness and 100% cotton yarn were also used as weft of the denims in order to investigate the differences statistically. It was found that that filament fineness and elastane draft had statistically significant effect on all inspected performances of denim fabrics except water absorption.

Keywords: denim fabric, dual core, multicomponent yarn, ring spinning system, microfilament

1. Introduction

In the clothing industry, denim has a wide acceptance with high potential uses as a fashion trend all over the world. Generally, denims are woven with a construction of 3/1 twill and they consist of indigo dyed warp yarns interlaced with gray weft yarns. Denim fabrics are rigid and durable. Denim market has great market size at a value of $57,312.5 million in 2016, and it is forecasted that the denim market will have an annual growth rate of 6.4% during the forecast period. The denim sector with largest contribution rate to the global revenue with higher preferable products among the teenagers and adults, will expected to continue in growing in the future jean market [1].

With so much preference, denim sector has to constantly evolve the fabric properties to meet consumer needs such as flexible, shape retention, low abrasion resistance and comfortable to wear in use at all times. High competitive potential in this sector pushes the companies to offer opportunities of different fiber, functional yarns use which contribute fabric properties. One of the best offered way which response to these requirements is using core-spun yarns. These yarns are produced by wrapping sheath fibers around filament or staple fiber core with a certain twist [2, 3]. Incorporating core part within the yarn structure makes the yarn cheaper, stronger and qualified, besides, sheath fibers are conserve the traditional appearance, handle and comfort properties. It is exemplified as using cotton covered elastic core-spun yarn is a good example to enable free movements and at the same time provide higher fabric comfort with cotton in the yarn structure [4].

In the literature, many researchers have been focused on elastane containing core-spun yarns in terms of elastane draft, elastane ratio, elastane linear density, elastic yarn positioning, twist factor etc. in order to obtain optimum yarn properties [2, 5–18]. When twist factor increases, it will affect the tensile properties of elastic cotton core-spun yarn positively [12, 18]. Elastane draft and ratio are important factors influencing the yarn's mechanical properties. Elastane ratio influences the tenacity and elongation at break of wrapped elastane core-spun yarns with the same twist factor. In that way, the core-yarn's tenacity and elongation decreased while the elastane percentage increased [8, 9, 18]. Elastane draft effects breaking tenacity and elongation [10]. On the other hand, higher draft ratio causes decrease in elastic recovery of elastic core spun yarn [6]. Proper tension control of elastane and use of a thread guide device helps to keep the spandex at the center of the yarn, thereby improving the quality of the elastic core-spun yarn [18]. Elastic yarn positioning in the yarn construction is another important parameter on yarn characteristic [5, 9].

Besides using of elastic core-spun yarn, filament core-spun yarns are also widely used in order to enhance some characteristics of fabric i.e. durability, esthetic and functional properties. Filaments used for filament core-spun yarns can vary such as polyester, polyamide, polybutylene terephthalate (PBT), T400®, polypropylene etc. Sarioglu and Babaarslan studied on physical properties of filament core-spun yarns having different filament fineness (conventional, fine and micro) and yarn linear density. False twist textured polyester filaments with different fineness were used and cotton covered polyester filament core-spun yarns were manufactured by modified ring spinning system. They concluded that filament fineness and yarn linear density had a great influence on yarn breaking strength and elongation. Higher core ratio was found to have higher strength and elongation. In addition, filament fineness had a significant effect on the unevenness properties and it was determined that there was no statistical effect on imperfections except for yarn linear density parameter. Thus, it can be said that filament core-spun yarns have got better performances in comparison to 100% combed cotton ring spun yarns except hairiness properties [19]. Erez and Çelik investigated the influence of both twist factor and filament blend ratio (core-sheath ratio) on properties and liveliness of cotton wrapped polyester filament core-spun yarns. They found that twist factor and core-sheath ratio had a significant effect on yarn strength, elongation, hairiness, liveliness and diameter [20]. Jeddi et al. studied both structural and physical properties of cotton covered nylon filament core-spun yarns with different twist factor and filament pretension [21]. Mahmood et al. stated that the best results for yarn strength of cotton covered nylon monofilament core-spun yarn were obtained at minimum twist factor and lowest spindle speed [3]. Shanbaz et al. studied on cotton covered polyester filament core-spun yarn on a modified ring frame to obtain the effectiveness of percentage of filaments in the blend, twist factor, positioning of roving on count, lea yarn strength and count strength product. Twist factor and roving positions have highly significant influence on count, lea strength and count strength product [22]. Similarly, Çelik et al. determined the influence of both twist factor and filament blend ratio on strength of yarn with the same materials. When sheath percentage increased, higher unevenness and lower hairiness were obtained [23]. Rameshkumar et al. investigated the core positioning on sheath coverage, core sheath ratios and plying effects on yarn and knitted fabric properties using polyester filament and waste silk. Tenacity, elongation and CVm of yarns also improved with increasing the core components. The core positioning at the center had lower tenacity with respect to right and left positioning. Thermal conductivities of silk rich (67% silk) fabrics were higher in comparison polyester rich (37% silk) fabrics. It was proved that polyester rich core-spun yarn fabrics show higher wicking [24]. Pramanik and Patil compared with cotton covered with crimped and drawn polyester filament hard-core ring and air jet yarns and 100% cotton ring yarn. It was concluded that using filament as core improved strength, elongation of filament core-spun yarns and ring spun and air jet spun yarns showed better performance with respect to 100% cotton ring-spun yarn [25]. Polyvinyl alcohol (PVA) is another filament used as tracer fibers in production of core-spun yarn. PVA is a water soluble filament, can be extracted from yarn structure by hot water easily. After extracted PVA filament from core-yarn, higher elongation, but similar breaking strength can be obtained when compared with typical ring yarns [26].

With technological progress and rising demands to obtain durable and long lifetime fabric, now the use of multicomponent core-spun (dual core-spun) yarns which enables elasticity and durable at the same time available. In other words, dual core-spun yarns consist of two core components; filament and elastane. In the production of the dual core-spun yarns filaments used can be polyester, T400®, polybutylene terephthalate (PBT) etc. In order to obtain dual core-spun yarns, additional creel loading of filament apparatus should be added to the current spinning system. As seen in the literature, there are lots of studies in terms of elastic core-spun and filament core-spun yarns with different materials.

Hua et al. developed elastic core-spun yarn containing a mix of spandex and polyethylene terephthalate/polytrimethylene terephthalate (PET/PTT) bi-component filament as core to obtain better yarn properties, especially for elastic property. In this study spandex draft ratio and linear density were selected as parameters and results showed that yarn stress decay, CVm value of evenness, and hairiness decrease when PET/PTT bi-component filament and spandex filament were used together as core. Furthermore, better yarn evenness was obtained by using PET/PTT bi-component filament and spandex filament core [27]. Telli et al. studied on fabrics containing tungsten in order to contribute alternative electromagnetic shielding. The fabric samples with three different yarns as core; Inox, Copper and Tungsten wires and three different double core-spun yarns with elastane and metal wires were produced. Electro Magnetic Shielding Effectiveness (EMSE) performances of fabrics were then evaluated [28].

With all these progresses show that the production of alternative functional fabrics has become possible with the use of dual core-spun yarns within the fabric structure. El-Tantawvy et al. investigated the pilling properties of fabrics produced from dual core-spun yarns with and without welding process. Elastic core-spun yarn was also produced to determine the differences. They found that both types of dual core yarns exhibit less pilling then the core spun yarn fabrics [29]. Bedez Ute was focused directly on mechanical and dimensional properties of denim fabrics made from double core and core-spun weft yarns used at different densities. It was concluded that weft density effect was higher than weft yarn composition for mechanical and dimensional properties of denim yarns [30].

It is envisaged that researches will be developed in the production of different yarn compositions, so the use of dual yarns in denim fabric production will probably become widespread due to its advantages properties with respect to conventional ones. This study was carried out in order to contribute to the use of dual core-spun yarns in denim fabrics and bring a different perspective. This is experimental study is designed in order to compare breaking force, breaking elongation, static tear force, elastic recovery, moisture management i.e. vertical wicking and water absorbency rate of twill 3/1 denim fabrics made from cotton covered filament, both filament and elastane core-spun yarns and 100% cotton yarn in weft.

2. Materials and methods

2.1. Materials

This study was conducted about the performance of denim fabrics containing dual core-spun yarns with filament fineness and elastane draft ratio variables. In that respect, the properties of drawn textured polyester filaments with conventional, fine and micro fineness are given in **Table 1**. All polyester (PET) filaments were selected among the most commonly used commercial types which are named as stretch textured yarns. Since microfilaments are sensitive to heat and it is necessary to omit the second heating zone during texturing, the PET filaments are in set form.

Basically, the study was focused on three different yarn types such as 100% cotton (Co) yarn, cotton covered filament core-spun yarn by using 110 dtex PET filament with different filament fineness as a control variables, and also cotton covered dual core-spun yarns by feeding 110 dtex PET filament with different filament fineness and 78 dtex elastane filament as core within the yarn structure. In the yarn production, cotton fiber with the physical properties of 28.53 mm staple length, 4.56 micronaire fineness, 31.53 gf/tex tenacity and 28.53% elongation was used as a sheath fiber.

2.1.1. Yarn and denim fabric production

Dual core-spun yarn samples were produced with modified ring spinning system which was designed by adding an extra creel for facilitating both elastane and filament feeding at the same time as a core. Schematic representation of dual core-spun yarn production, combination of materials and cross-sectional view of yarn are illustrated in **Figure 1(a–c)**, respectively.

http://dx.doi.org/10.5772/intechopen.80286

As seen from this **Figure 1**, both PET filament and elastane are driven by positive feed roller, separately. These components are fed to the nip point of the front rollers by means of V-grooved roller and at the same time cotton fiber wraps over these components, as well (**Figure 1b**). Thus, the draft of PET filament and elastane are achieved by speed difference between yarn delivery and front roller of drafting unit. Here, PET filament draft was kept constant as 1.08. On the other hand, draft of elastane was varied with 2.9, 3.2, 3.5 and 3.8. In doing so, 16 Co/PET/Elastane dual core-spun yarn samples were obtained with two different cores in order to benefit from these properties.

Parameters	Conventional	Fine		Micro
Linear density (dtex)	110	110	110	110
Number of Filament	36	96	192	333
Tenacity (cN/dtex)	4.07	3.90	3.45	3.80
Elongation (%)	22.65	23.66	21.85	26.84
Crimp Stability (%)	85.93	84.08	78.50	77.32
OPU (%)	1.16	3.02	2.24	2.52
Intermingling (number/m)	52.40	76.10	73	67.60

Table 1. PET filaments properties.

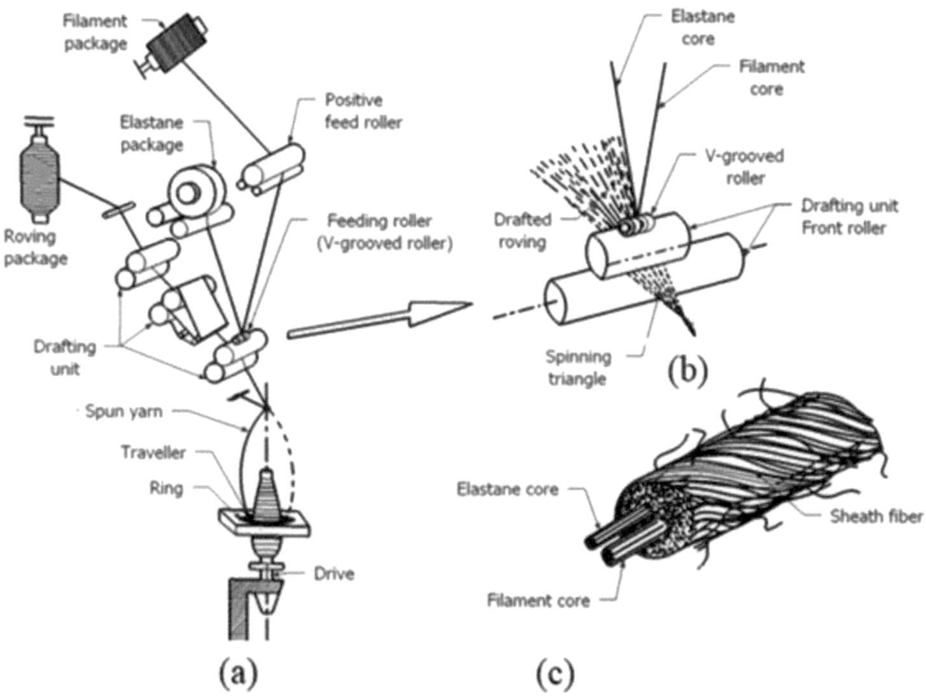

Figure 1. Schematic illustration of modified ring spinning system, positioning the PET/elastane core at the nip point of the front roller, double core-spun yarn view (It may not be reproduced without permission); (a) modified ring spinning frame, (b) combination of materials, (c) simulated longitudinal and cross-sectional view of double core-spun yarn containing filaments [32].

In the production of the PET filament core-spun yarns, same system was used without extra elastane feeding and all production parameters were kept constant. Moreover, 100% Co ring spun yarn was also produced without both PET and elastane feeding. Combed cotton roving with 844 tex linear density was used for the production of all yarn types in order to produce 42 tex yarn samples at 9500 rev/min spindle speed and 660 turns/m twist.

These yarn samples were used as weft yarn and indigo dyed 100% Co ring spun yarn with 59 tex linear density was used as a warp yarn. Twill denim fabrics with 3/1 pattern were manufactured at constant structure parameters such as; 26 ends/cm warp density, 20 picks/cm weft density, 480 rev/min weaving machine speed, 180 cm reed width, 65/4 reed number. After the denim fabric production, singeing, desizing and finishing processes and thermal fixation processes were carried out. Finally, 21 different denim fabric samples were obtained. Design of experiment for denim fabrics is shown in **Table 2**.

2.2. Methods

Denim fabric samples were conditioned in standard atmosphere conditions at 20 ± 2°C temperature and 65 ± 4% relative humidity for 24 hours in accordance with BS EN ISO 139 standard [33]. Tested denim fabric properties, related standards and test proceduresare illustrated in **Table 3**.

Fabric type	Filament fineness (dtex)	Elastane draft
100% Co denim	—	—
Filament core-spun denim	3.05, 1.15, 0.57 and 0.33	—
Dual core-spun denim	3.05, 1.15, 0.57 and 0.33	2.9, 3.2, 3.5 and 3.8

Table 2. Design of experiment for denim fabric samples.

Properties	Standard	Procedure
Fabric weight	ISO 3801	100 cm^2 of each fabric were cut with dies and weighed on a precision scale and then multiplied by 100. The fabric weight was calculated in g/m^2.
Fabric thickness	ISO 5084	Thickness measurements for each fabric samples were taken by means of digital thickness tester.
Warp and weft density	BS EN 1049-2	The numbers of warp and weft yarns in 1 cm were determined for each fabric samples.
Breaking force and elongation	BS EN ISO 13934-1	Breaking force and elongation for both warp and weft direction were determined at 200 mm gauge length, 100 mm/min test speed.
Static tear force	BS EN ISO 13937-2	Test samples were prepared in accordance with single-tear method. Tests were performed for warp and weft yarns, separately, for each specimen at 100 mm/min test speed.
Elasticity	BS EN 14704-1	To determine the elasticity properties of fabrics, Method A was used. The number of the cycling load was 50 cycles (instead of 5 cycles because of detecting elastic recovery under higher number of cyclic loading) with applying load of 6 N/cm width of the fabric.

Table 3. Test standards used and procedure achieve to determine denim fabric properties.

2.2.1. Moisture management

"Moisture Management" includes all the terms of wicking, wetting, absorbency or transportation and these properties of the fabrics are related to the ability of a textile fabric to transport moisture away from the skin to fabric's outer surface in multi-dimensions. Wetting and wicking are considered as the most important parameters for absorption and transportation of liquid in textile clothing [31].

Wicking is the flow of a liquid in a porous substance in time which is driven by capillary forces [32]. Vertical wicking rate was conducted by various researchers in different ways [31–35]. In this study, to evaluate vertical wicking properties of denim fabrics, 20 cm × 2.5 cm strip test specimens for warp and weft direction were prepared. Denim fabrics were suspended vertically with its 3 cm of lower end immersed in a reservoir of distilled water colored with 0.01 g red dye to observe the rate of the uptake of the liquid easily. Because of the high areal density of denim fabric samples, to be tested pretension was applied with two clips totally to ensure 2 g of dead load. The wicking height of liquid rising was measured after 30 min time intervals and wicking rate was determined as mm/s. Wicking tests were conducted with five samples for both warp and weft directions of each fabric. In order to complete the wicking tests more quickly, a new wicking apparatus design was made (**Figure 2**). The designed apparatus consists of five clamps for sample hanging, five rulers placed next to the fabric samples for height measurement and reservoir. Since it is planned that five specimens will be applied to vertical wicking test procedure for each sample, such an apparatus is desinged and manufactured in order to comple the wicking tests in shorter time.

Wettability terms is explained as the first impression of fabric when get into touch with liquid. When the fabric is wetted the interaction between the forces of cohesion (within the liquid) and the forces of adhesion (between the fibers and the liquid) determines whether wetting takes place or not and also determines spreading and absorption of the liquid over the surface of the textile material [31]. In order to determine the absorption areas of the fabrics after 0.2 ml water was dropped, image analysis was conducted by using camera to catch the visual after 2 min.

In order to determine the absorption rate and area of the fabrics wetted, image processing method was used. For this aim, an image acquisition system was built up. The system consists of a digital microscope camera, lightening system, camera attachment equipment and computer (**Figure 3**). Since the liquid existence within the fabric structure will lead to different light transmission level in comparison to dry regions, it was considered that the dry and wetted regions can be distinguished by applying a logical threshold operation in accordance with the pixel light intensity values. So, the back lightening system was selected for this study.

The denim fabric samples were placed over the lightening unit. The colored solution with 0.2 ml volume was dropped on the fabric sample by means of a screwed syringe. At the same time, the video acquisition of the digital microscope camera was started. The image frames with the size of 640 × 480 pixels were snap shotted at instant of solution drop fell on the sample and 2 min after dropping. These image frames in JPG format were acquired from three different parts of each fabric sample. The acquired image frames were analyzed by means of a developed algorithm (**Figure 4**).

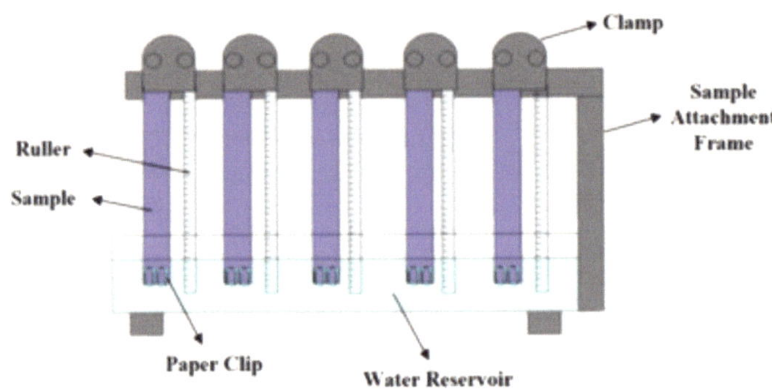

Figure 2. Vertical wicking apparatus.

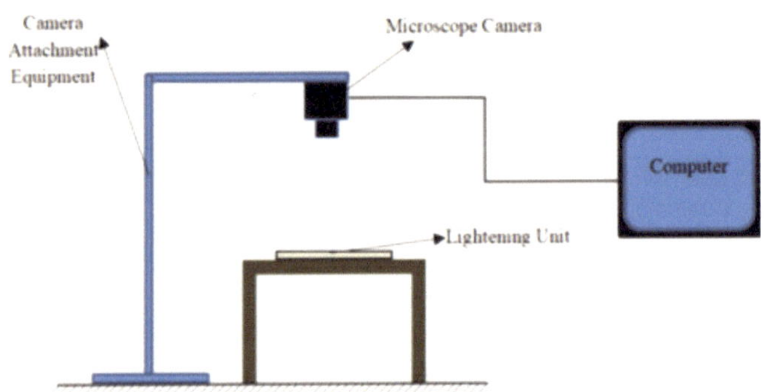

Figure 3. Image acquisition system.

First of all, the color image frame in RGB format was converted into gray scale. Then, the image enhancement operations such as noise removing and smoothing were applied by using average and Gaussian filters respectively. Gaussian smoothing commonly forms the first stage of an edge-detection algorithm [36]. The average filter is useful especially for removing Gaussian noises. In order to eliminate the noises caused from lightening condition and electrical reasons, the noise removing filters are applied [36]. The enhanced image frame was converted into binary form by applying a suitable threshold value. All pixel values of the filtered gray image were replaced with the value 1 (white) when the corresponding pixel value greater than threshold level, otherwise it was replaced with the value 0 (black). The white pixels in the binary image correspond to the liquid absorbed area and the black pixels correspond to dry area. In order to determine the exact absorption area and remove the other unnecessary parts, opening morphological operation was applied. Morphological opening is erosion followed by dilation, using the same structuring element for both operations. The opening operation has the effect of removing objects that cannot completely contain the structuring element. Boundary of the absorption area was labeled by means of "canny" edge detection method. Finally, the area of the absorption region was calculated. The absorption rate of each denim fabric sample is calculated as the percentage of white pixels to the whole pixels of the binary image. The application results of the developed algorithm at the instant of drop fell and 2 min after dropping were given in **Figures 5** and **6**, respectively.

http://dx.doi.org/10.5772/intechopen.80286

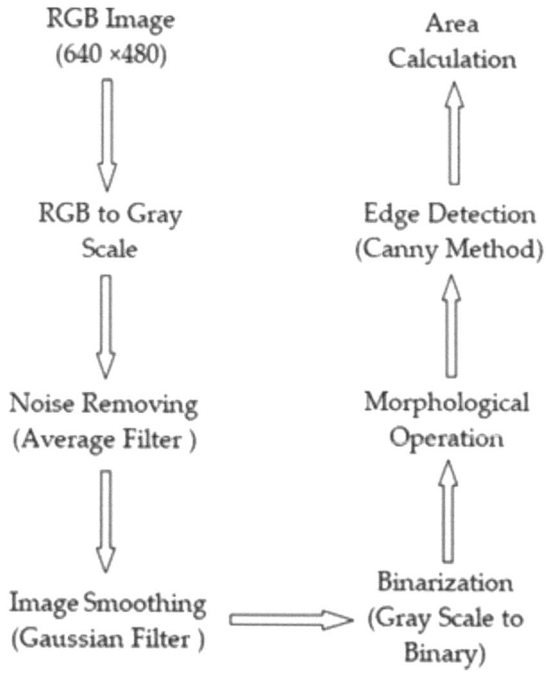

Figure 4. Absorption area calculation algorithm.

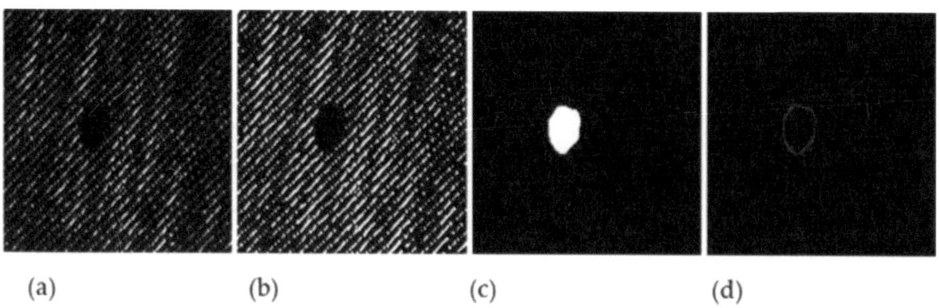

Figure 5. Absorption area at the drop fell.

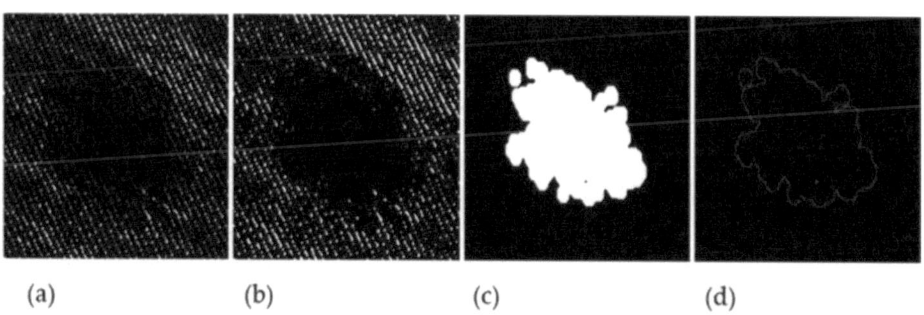

Figure 6. Absorption area 2 min after dropping.

2.2.2. Statistical analysis

In statistical analysis, multivariate analysis of variance (MANOVA) was achieved at 95% confidence interval by means of SPSS package program to determine whether there was statistically significant effect of the filament fineness, and elastane draft on denim fabric breaking force, breaking elongation, static tear force, elasticity, wicking rate and water absorption rate. The evaluated independent parameters were used as weft yarn in the denim fabric production so all response variables were analyzed in weft direction as well.

3. Experimental results

The structural properties of denim fabrics and test results are illustrated in **Tables 4** and **5**, respectively.

Fabric type	Filament fineness (dtex)	Elastane draft	Fabric weight (g/m²)	Fabric thickness (mm)	Warp density (ends/cm)	Weft density (picks/cm)
100% Co denim	—	—	310	0.68	29.6	22
Co/PET	3.05	—	320	0.71	29.8	22.4
Filament core-spun denim	1.15	—	320	0.74	29.6	22.8
	0.57	—	320	0.68	29.8	22.8
	0.33	—	330	0.70	29.8	22.4
Co/PET/elastane	3.05	2.9	342	0.83	32.4	22.8
Dual core-spun denim	1.15	2.9	355	0.81	32.6	22.8
	0.57	2.9	345	0.83	33	22.8
	0.33	2.9	352	0.79	32.4	22.8
	3.05	3.2	352	0.83	30.6	22.8
	1.15	3.2	351	0.80	32.8	22.8
	0.57	3.2	350	0.81	32.8	22.8
	0.33	3.2	357	0.83	32.6	23.2
	3.05	3.5	355	0.84	33.4	22.8
	1.15	3.5	360	0.83	33	22.8
	0.57	3.5	359	0.84	33	22.8
	0.33	3.5	360	0.84	33	22.8
	3.05	3.8	356	0.83	30.8	22.8
	1.15	3.8	354	0.84	33.4	22.8
	0.57	3.8	359	0.82	33.4	22.8
	0.33	3.8	365	0.82	33.6	22.8

Table 4. Structural properties of denim fabric samples.

3.1. Breaking force and elongation

To advance the comfort performance of the denim fabrics during body movement, dual core-spun yarns including elastane that provide higher elasticity and recovery are preferred. However, this advantage brings a disadvantage together and it leads to decrease in tensile strength of denim fabrics [30, 37]. The tensile test outcomes of the presented study is illustrated in **Figure 7** the breaking force of denim fabrics in weft wise composed of 100% Co, filament core-spun yarns and dual core-spun yarns.

It was clearly seen in **Figure 7** that denim fabrics made from filament core-spun yarns have the higher breaking forces than both 100% Co and dual core-spun denims. The highest breaking force was detected at 0.33 dtex filament core-spun denim fabric, this result is attributed to the fact that more filaments in the yarn cross-section can provide more resistance against tensile force. When the breaking forces of dual core-spun denim fabrics are taken into consideration in terms of effect of elastane draft, it is observed that the breaking force of samples with conventional firmament have increasing trend by increasing elastane draft. However, the fabric samples with micro fineness filament have decreasing trend with the elastane draft increase. Since increasing the elastane draft leads to decrease of elastane ratio within the fabric increase and so increase in breaking force of the fabric. However, this result is not clearly seen for dual core-spun denim fabrics due to involving both PET filament and elastane.

The breaking elongation of denim fabrics are shown in **Figure 8**. Elastane content contributes the elongation of the denim fabrics and this situation is clearly observed among the all fabric samples. Elongation directly affects the elasticity properties of the denim fabrics this is why elastane is used. The lowest breaking elongation was obtained with 100% Co denim fabric without both filament and elastane. When the denim fabrics made from filament core-spun yarns are investigated, it is seen that 3.05 dtex filament core-spun denim fabrics has the highest breaking elongation value and breaking elongation values of all filament core-spun denim fabrics are similar. From these results, it can be revealed that filament core part contributes strength of the fabric with an acceptable elongation value.

It can be possible to see how dominant the effect on the denim fabric elongation performance of the elastane draft ratio is. It is observed that with the increase of the elastic draft ratio, the breaking elongation of the fabric increases. This situation may be attributed to the fact that higher elastane draft may probably leads to increase in cohesion forces between filament, elastane and Co, and so breaking elongation can probably increase. Hereby, it can be said that elastane represents a large majority of the extensible part of fabric under the tensile force. In terms of filament fineness parameter of dual core-spun denim fabrics, it can be said that breaking elongation varies from 38.18 to 44.08%. The predominant elasticity property of elastane makes it difficult to see the effect of filament fineness in **Figure 8**, however, with statistical analysis effects can be examined in detail.

3.2. Static tear force

Static tear force in weft direction of denim fabric samples are illustrated in **Figure 9**.

According to tear force values, it is seen that 100% Co denim fabric has lower value than that of denim fabrics containing both filament and dual core-spun yarns. On the other hand, PET

Fabric type	Filament fineness (dtex)	Elastane draft	Breaking force (N)		Breaking elongation (%)		Static tear force (N)		Elastic recovery (%)		Vertical wicking rate (mm/s)		Water absorption (%)
			Warp	Weft	Warp	Weft	Warp	Weft	Warp	Weft	Warp	Weft	—
100% Co denim	—	—	1321.89	753.86	25.70	21.44	31.85	35.63	35.85	56.63	0.0496	0.0417	10.14
Co/PET	3.05	—	1372.94	864.41	26.16	26.08	56.96	47.29	41.98	78.02	0.0524	0.0500	23.72
Filament core-spun denim	1.15	—	1403.24	831.55	24.23	24.77	57.60	47.86	47.50	77.68	0.0419	0.0469	21.28
	0.57	—	1441.81	832.33	28.51	24.33	57.82	47.78	47.48	80.84	0.0317	0.0472	21.30
	0.33	—	1335.22	888.84	26.98	25.45	56.33	50.50	43.33	80.34	0.0498	0.0444	22.16
Co/PET/elastane	3.05	2.9	1604.25	739.35	32.21	39.39	56.90	42.93	51.61	91.62	0.0506	0.0511	21.98
Dual core-spun denim	1.15	2.9	1533.54	781.09	32.15	38.18	55.42	48.84	52.55	92.37	0.0457	0.0509	26.30
	0.57	2.9	1419.95	809.82	30.95	38.43	55.70	48.46	55.51	92.40	0.0467	0.0502	22.15
	0.33	2.9	1581.12	799.83	32.96	38.67	55.92	49.44	53.03	87.18	0.0481	0.0485	29.24
	3.05	3.2	1493.31	764.48	31.05	40.96	55.50	45.68	54.31	88.35	0.0480	0.0489	20.04
	1.15	3.2	1327.06	801.72	31.17	41.49	54.43	49.58	55.09	87.97	0.0426	0.0474	25.01
	0.57	3.2	1529.27	801.60	32.22	41.37	55.20	50.70	55.47	89.82	0.0333	0.0439	19.23
	0.33	3.2	1586.84	801.25	33.42	41.42	55.36	48.70	52.58	89.51	0.0385	0.0480	26.18
	3.05	3.5	1606.22	763.32	32.59	41.81	55.60	45.27	55.29	86.88	0.0487	0.0463	16.67
	1.15	3.5	1658.09	778.34	32.87	41.44	53.77	47.73	52.76	90.17	0.0472	0.0463	24.92
	0.57	3.5	1465.46	794.11	31.46	42.81	54.44	46.64	52.70	89.59	0.0461	0.0433	20.72
	0.33	3.5	1632.86	777.56	32.33	41.67	55.07	46.27	53.75	90.76	0.0411	0.0457	29.30
	3.05	3.8	1611.52	760.10	31.48	44.08	55.93	44.28	39.53	89.85	0.0365	0.0478	17.34
	1.15	3.8	1547.79	792.34	32.23	42.47	54.81	48.49	46.89	91.24	0.0448	0.0485	28.47
	0.57	3.8	1572.50	779.58	31.80	44.00	53.35	46.64	50.12	90.82	0.0300	0.0507	21.13
	0.33	3.8	1574.92	784.28	31.58	43.23	53.96	47.39	50.40	90.38	0.0489	0.0474	28.07

Table 5. Test results of denim fabric samples.

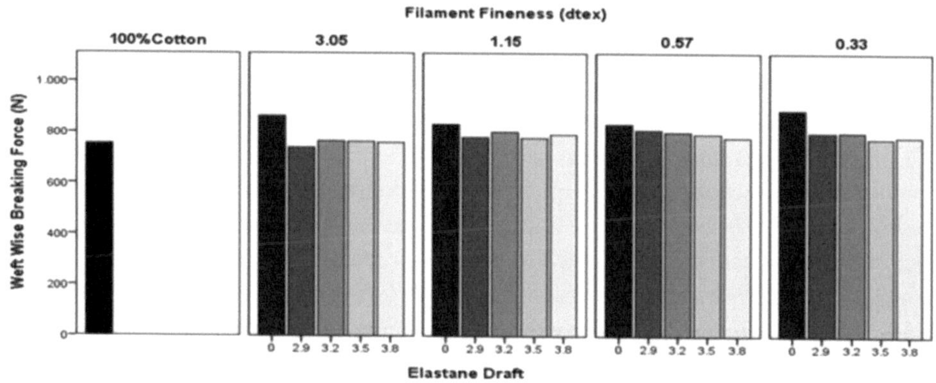

Figure 7. Weft wise breaking force of denim fabrics.

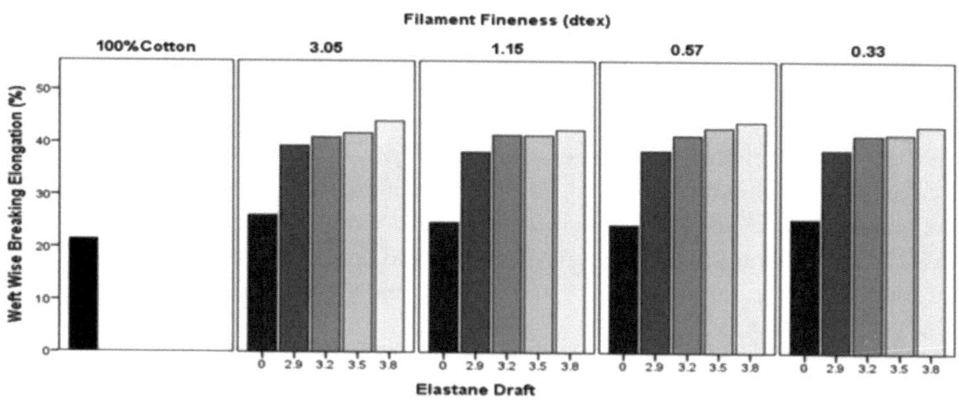

Figure 8. Weft wise breaking elongation of denim fabrics.

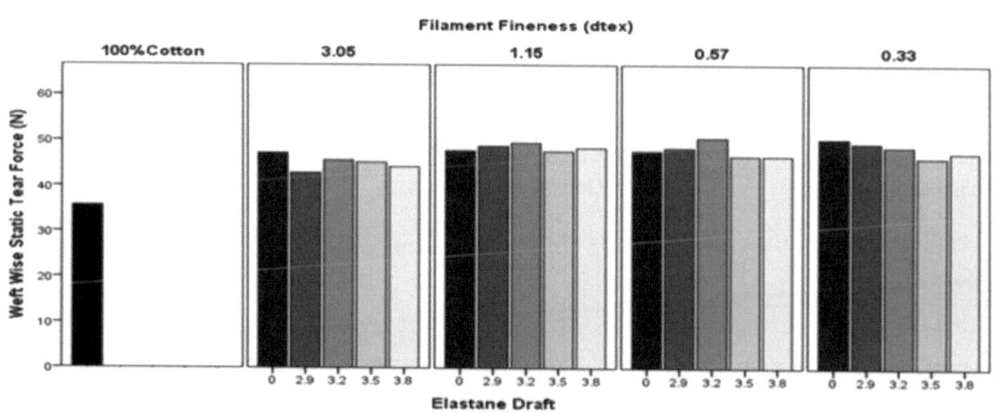

Figure 9. Weft wise static tear force of denim fabrics.

filaments from conventional to micro fineness contribute static tear force of denim fabrics except 0.57 dtex microfilament containing filament core-spun denim fabric. PET filament with 0.57 dtex filament fineness has the lowest breaking strength and so it can be said that filament properties affect the denim fabric properties as well. In addition, static tear force increases until 3.2 elastane draft ratio for all dual core-spun denim fabric types except 0.33 dtex filament containing dual core-spun denim fabric. Then, it is observed that after 3.2 draft value the increase in elastane draft effects static tear force negatively. Static tearing action leads to as the broken of the yarns individually or group. Hereby, increasing the elastane draft contributes tear force until 3.2 draft value because of the rising Co content. Furthermore, PET filament with fine and micro fineness contribute the tearing performance of the denim fabrics, because of the higher number of filament in the core of the dual core-spun yarns acts more resistance to break. This result can be explained with higher number of filament in the core of the dual core-spun yarns providing more resistance to break.

3.3. Elastic recovery

During usage of the denim should stretch freely in accordance with body movements especially at knee and should retain its original shape without any deformation after stretching. So that capability of extension and recovery of denim after repeated loadings is very important characteristics [38]. Higher the number of the loadings can contribute to life assessment in accordance with evaluating the fabric performance. Different from the test standard, 50 cycle loadings were carried out in order to evaluate elastic recovery of denim samples rather than 5 of cyclic loadings. Elongation after 50 cyclic loading and un-recovered elongation after 60 min recovery period were estimated according to Eqs. (1) and (2), respectively in accordance with BS EN 14704-1. Elongation, *S*, expressed as percentage:

$$S = \frac{E - L}{L} * 100 \tag{1}$$

where: E = extension (mm) at maximum force on the 50th cycles, L = initial length (mm).

Un-recovered elongation, *C*, expressed as percentage:

$$C = \frac{Q - P}{P} * 100 \tag{2}$$

where: Q = distance between applied reference marks (mm) after a specified recovery period, P = initial distance between applied reference marks (mm).

From Eqs. (1) and (2) elastic recovery of fabric can be expressed as following Eq. (3).

$$R = \frac{S - C}{S} * 100 \tag{3}$$

Figure 10 displays weft wise elastic recovery of denim fabrics after waiting 60 min recovery period. It can be said that both filament and elastane improve stretchability and recovery properties of the denim fabrics when compared with pure Co ones.

In terms of elastic recovery value of denim fabrics, pure Co denim has the lowest stretchability properties. The presence of elastane contributes to elastic recovery with a high value approximately 85–90%. Dual core-spun denim fabrics have also higher elastic recovery than filament core-spun fabrics. In general it can be said that increase in elastane draft also increases the elastic recovery except 2.9 elastane draft. It is seen that denim fabrics have the highest elastic recovery at 2.9 elastane draft of 3.05, 1.15 and 0.57 dtex dual core-spun yarns.

3.4. Wicking rate

The wicking rate measurements in mm/s are presented in **Figure 11**. Weft wise wicking rate obtained for pure Co denim is lower than vertical filament core-spun denims. Vertical wicking rate of the filament core-spun denim fabrics decreases from 3.05 to 0.33 dtex. Since filaments are in the core part of the yarn, capillary transfer of water may not be fully observed during the period of time. This situation can be observed at the rest of the denim fabrics with different draft ratio. On the other hand, in general, the effect of the elastane draft can be seen

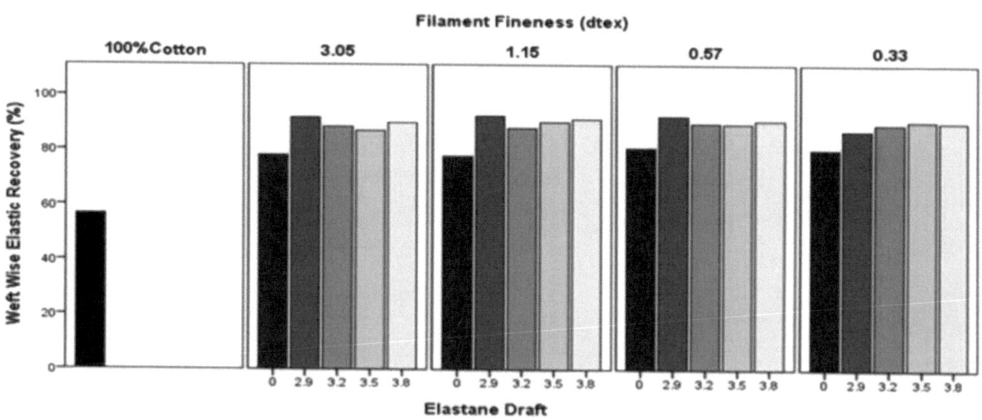

Figure 10. Weft wise elastic recovery of denim fabrics.

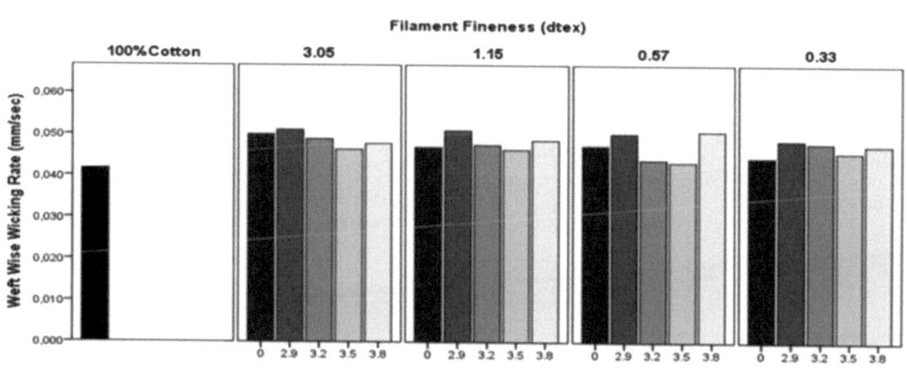

Figure 11. Weft wise wicking rate of denim fabrics.

as decreasing wicking rate of the denim fabrics. Decreasing in elastane draft causes increase in Co and it is known that Co absorbs the water instead of transfer, as well. In addition, it is also observed that increase in the number of filaments causes a decreasing wicking rate when the average of the values for each filament fineness including all elastane draft is taken into consideration.

3.5. Water absorption rate after dropping

Water absorption rate as percentage of the wetted area detected by image processing method by using MATLAB is illustrated in **Figure 12**. Naturally, it is known that the absorption of cotton fibers is better when compared to synthetic fibers. Water transfer occurs in synthetic fibers through capillary forces. So, the rate of absorption in pure Co denim i.e. the area of dropped colored water is found smaller after 2 min. On the other hand, it is seen that filament core-spun denim fabrics have at least two times greater absorption area than that of pure Co fabric. It can be said that the presence of PET filament leads to water or moisture transportation instead of penetration. It can be observed that there is no increase or decrease tendency of absorption rate of dual core-spun denim fabrics in terms of elastane draft. This situation can be explained with having different liquid transportation of fabrics depending on the filament fineness. Whereas the elastane has taken almost the entire stretch of fabrics, the PET filament here carries capillary properties of fabrics with its high capillary transport capability in moisture management.

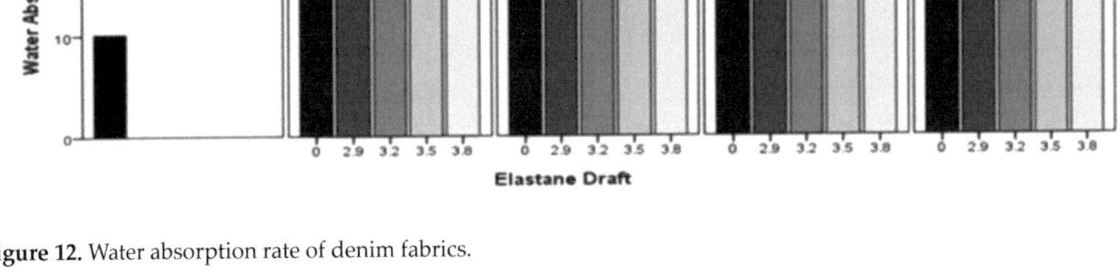

Figure 12. Water absorption rate of denim fabrics.

Independent variables	Breaking force (N)	Breaking elongation (%)	Static tear force (N)	Vertical wicking rate (%)	Elastic recovery (%)	Water absorption rate (%)
Filament fineness	0.000*	0.003*	0.000*	0.000*	0.000*	0.000*
Elastane draft	0.000*	0.000*	0.000*	0.000*	0.000*	0.077

*The mean difference is significant at the *0.05* level.

Table 6. Multivariate analysis of variance (MANOVA) test results for weft wise properties of denim fabrics.

In analyzing of filament fineness effect, 3.05 and 0.57 dtex dual core-spun yarn denim fabrics including all elastane draft have the lower wetted area percentage with the average value of 19 and 20.8%, respectively. On the other hand, 1.15 and 0.33 dtex dual core-spun yarn denim fabrics including all elastane draft have higher wetted are percentage with the average of 26.2 and 28.2%, respectively. The highest wetted area was found as 0.33 dtex, it can be explained as higher number of the filaments enables more liquid transportation.

3.6. Statistical analysis

To put forward the influence of filament fineness and elastane draft parameters on denim fabric breaking force, breaking elongation, tear force, elastic recovery, vertical wicking rate and water absorption rate statistically, multivariate analysis of variance (MANOVA) was achieved at 95% confidence interval (**Table 6**). Statistical results indicate that filament fineness parameter has significant effect on all variables. Elastane draft has also significant importance on all depended variables except water absorption rate ($p = 0.077$) at the *0.05* level.

4. Conclusion

Application of core-spun yarns in textile industry is in progress to improve physical and mechanical properties of fabrics, such as comfort, abrasion resistance, tenacity, durability, and functional properties. Stretch denim fabrics are mostly produced from core-spun yarns. Advanced stretching performance provides better fitting to body. Cotton is the most appreciable material as sheath component of the yarn which is responsible for comfort and esthetic properties. Denims' clothing and fitting to human body, comfortable and performance properties are essential for consumers. The most acceptable driving factors for denim market are consumer demand, rapid change of fashion and denim styles, highly preferred by young people and these factors are changing in very short period of time. When the overall consumption of denim in the world and market size are taken into consideration, value-added and high durable denims produced from functional yarns will response to the desired comfort and fit characteristics as well.

Since the denim fabrics have high demand in textile market and it is increasing day by day, this study is conducted to evaluate the developments in denim fabric production and submit an innovative case study to improve the performance of denim fabric. In day fashion trends, denim fabrics are now not only used for the jeans production, they are also used in the production of shirt, t-shirts, skirts, bags and different textile product accessories. Depending on the usage area of denim fabrics, different performances and functions are required from them. Different ways can be performed to improve the performances of the textile products. Using different pattern designs, selecting proper raw material, using specially produced yarns or applying finishing chemicals are effective treatments. The finishing and washing processes are applied to add higher hand property, better touch feeling and attractive appearance. Evidence in literature demonstrates that usage of different characteristic fiber and yarn is effective method to increase mechanical performance of the denim fabrics. The incorporation of elastane and filament in yarn core gives new configurations to yarn geometry and ultimately changes the fabric geometry. Depending on developments in the yarn production technology, specially designed and functional yarns such as core-spun, filament core-spun and dual core-spun yarns are produced and widely used.

In this study, to design more attractive and higher performance denim fabric Co/PET/elastane containing dual core-spun yarns were systematically used in production. Furthermore, the effects of filament fineness and elastane draft on denim fabric performance were revealed. Filament core-spun yarns with different filament fineness and pure Co yarn were also produced to compare the performance of denim fabrics. These all yarn types were used as filling of denims. Breaking force, breaking elongation, static tear force, elastic recovery, vertical wicking rate and water absorption rate properties were inspected and results were analyzed statistically.

As expected from the structural properties of the core-spun yarns, incorporating filament contributes to breaking force of the denim fabric. On the other hand, it was observed that increasing elastane draft affects breaking elongation positively. This result is attributed to the fact that higher elastane draft leads to increase in cohesion forces between filament, elastane and cotton sheath. When dual core-spun denim fabrics containing PET filament with fine and micro fineness is considered, tear force raises from fine to micro fineness because of the higher number of filament in the core of the dual core-spun yarns and so higher resistance to break.

In order to propound the advantages of the core-spun yarn usage, the performance of pure cotton denim fabrics were compared. It was seen that denim fabrics from pure cotton has the lowest elastic recovery. These results exactly proved that core-spun yarns can improve the mechanical performance of the denim fabrics without sacrificing the good softness and feeling of cotton. When the effect of core-spun yarn structural parameters on denim elastic recovery performance were investigated, in relation to the statistical analysis, it was also presented that both filament fineness and elastane draft have significant effect. Elastic recovery of the fabric can be aligned as cotton denims < filament core-spun denims < dual core-spun denims. The results in this study also indicates that the core-spun yarns not only increases the mechanical performance, but they also advance the comport performance. Higher wicking and absorption rates were obtained with core-spun and dual core-spun yarn denims in comparison to pure cotton denim.

For the absorption rate calculation, a novel method based on image processing was developed. Thus, the absorption area calculation was achieved accurately. In this experience, it was determined that the irregular color hues of indigo dyed warp yarns made the measurement difficult. High fabric thread density also restricted this measurement. So, it is planned that the absorption rate of the fabric samples will be determined by using different light source and lighting condition in further study.

Acknowledgements

Authors would like to thank to Çalık Denim Textile Industry and Trade Inc. Research and Development Center for their contribution in the yarn and denim fabric samples production.

http://dx.doi.org/10.5772/intechopen.80286

Author details

Osman Babaarslan[1], Esin Sarioğlu[2*], Halil İbrahim Çelik[3] and Münevver Artek Avci[4]

*Address all correspondence to: sarioglu@gantep.edu.tr

1 Department of Textile Engineering, Faculty of Engineering and Architecture, Çukurova University, Adana, Turkey

2 Department of Fashion and Textile Design, Faculty of Fine Arts, Gaziantep University, Gaziantep, Turkey

3 Department of Textile Engineering, Faculty of Engineering, Gaziantep University, Gaziantep, Turkey

4 Çalık Denim Textile Industry and Trade Inc. Research and Development Center, Malatya, Turkey

References

[1] Denim Market by Product (Jeans, Jackets and Shirts, Dresses), by Segment (Mass Market, Standard, Premium), by Consumer Type (Men, Women, Children), by Distribution Channel (Specialty Stores, Department Stores, Hypermarkets & Supermarkets), by Geography (U.S., Japan, Brazil, Germany, France, U.K., Italy, Canada, Mexico, Spain, Philippines, S. Korea, Australia, Pakistan, Russia, New Zealand, Turkey, Netherlands, South Africa, Thailand, Belgium, Malaysia)—Global Market Size, Share, Development, Growth and Demand Forecast, 2013-2023 [Internet]. 2018. Available from: https://www.psmarket-research.com/market-analysis/denim-jeans-market [Accessed: Mar 3, 2018]

[2] Shi F-J, Xuling J. Modelling the tensile properties of modal/polyurethane core spun stretch yarn. Fibres & Textıles in Eastern Europe. 2012;**20**:30-32

[3] Mahmood N, Jamil NA, Nadeem M, Saeed MA. Effect of multiple spinning variables on the spinability of cotton covered nylon filament core yarn. Pakistan Textile Journal; 2003;**LII**(2). [Internet]. Available from: http://www.ptj.com.pk/Web%202003/2-2003/art-nasir.htm [Accessed: 2017-10-08]

[4] Sinha SK, Bansal P, Maity S. Tensile and elastic performance of cotton/lycra core spun denim yarn. Journal of the Institution of Engineers (India): Series E. 2017;**98**:71-78

[5] Babaarslan O. Method of producing a polyester/viscose core-spun yarn containing spandex using a modified ring spinning frame. Textile Research Journal. 2001;**71**:367-371

[6] Su CI, Maa MC, Yang HY. Structure and performance of elastic core-spun yarn. Textile Research Journal. 2004;**74**:607-610

[7] Lou CW, Chang CW, Lin JH, Lei CH, Hsing WH. Production of a polyester core-spun yarn with spandex using a multisection drawing frame and a ring spinning frame. Textile Research Journal. 2005;**75**:395-401

[8] Dhouib AB, El-Ghezal S, Cheikhrouhou M. A study of the impact of elastane ratio on mechanical properties of cotton wrapped elastane-core spun yarns. Journal of the Textile Institute. 2006;**97**:167-172. DOI: 10.1533/joti.2005.0121

[9] Kakvan A, Najar SS, Saidi RG, Namil M. Effects of draw ratio and elastic core yarn positioning on physical properties of elastic wool/polyester core-spun ring yarns. Journal of the Textile Institute. 2007;**98**:57-63. DOI: 10.1533/joti.2005.0194

[10] Helali H, Dhouib AB, Msahli S, Cheikhrouhou M. Influence of dorlastan® draft and yarn count on elastic recovery of the dorlastan® core spun yarns following cyclic test. The Journal of the Textile Institute. 2012;**103**:378-384. DOI: 10.1080/00405000.2011.580542

[11] Das A, Chakraborty R. Studies on elastane-cotton core-spun yarns and fabrics: Part i: Yarn characteristics. Indian Journal of Fibre& Textile Research. 2013;**38**:237-243

[12] Shanbaz B, Nawaz SM, Saeed MA. Technological studies on elastic core cotton covered yarn. Pakistan Journal of Applied Sciences. 2002;**2**:690-693

[13] Singha K. Analysis of spandex/cotton elastomeric properties: Spinning and applications. International Journal of Composite Materials. 2012;**2**:11-16

[14] Qadir MB, Hussain T, Malik M, Ahmad F, Jeong SH. Effect of elastane linear density and draft ratio on the physical and mechanical properties of core-spun cotton yarns. The Journal of the Textile Institute. 2014;**105**:753-759. DOI: 10.1080/00405000.2013.848045

[15] Vuruşkan D, Babaarslan O, Ilhan I. Effect of production parameters on strength and elongation of the selected yarns containing elastane. Tekstil ve Konfeksiyon. 2011;**21**:22-29

[16] Pourahmad A, Johari MS. Comparison of the properties of ring, solo, and siro core-spun yarns. The Journal of the Textile Institute. 2011;**102**:540-547. DOI: 10.1080/00405000.2010.498170

[17] Kaynak HK. Optimization of stretch and recovery properties of woven stretch fabrics. Textile Research Journal. 2017;**87**:582-592. DOI: 10.1177/0040517516632480

[18] Su CI, Yang HY. Structure and elasticity of fine elastomeric yarns. Textile Research Journal. 2014;**74**:1041-1044

[19] Sarıoğlu E, Babaarslan O. A study on physical properties of microfilament composite yarns. Journal of Engineered Fibers and Fabrics. 2016;**11**:90-98

[20] Erez E, Çelik P. A research on the parameters of the affecting yarn properties of cotton-polyester rigid core-spun yarns. Tekstil ve Konfeksiyon. 2014;**24**:195-201

[21] Jeddi AA, Johari MS, Merati AA. A study of the structural and physical properties of cotton-covered nylon filament core-spun yarns. The Journal of Textile Institute. 1997;**88**:12-20. DOI: 10.1080/00405009708658525

[22] Shanbaz B, Nawaz SM, Ali R. Spinning performance and yarn properties of multiple filament polyester/cotton core yarn. Pakistan Journal of Applied Sciences. 2002;**2**:324-326

[23] Çelik P, Üte TB, Özden D, Çömlekçi H, Akkale EC. The effect of core/sheath ratio and twist amount on yarn characteristics of filament core yarns. Electronic Journal of Textile Technologies. 2009;**3**:29-37

[24] Rameshkumar C, Rengasamy RS, Anbumani N. Studies on polyester/waste silk core-spun yarns and fabrics. Journal of Industrial Textiles. 2009;**38**:191-203

[25] Pramanik P, Patil VM. Physical characteristics of cotton/polyester core spun yarn made using ring and air-jet systems. AUTEX Research Journal. 2009;**9**:14-19

[26] Moghassem AR, Gharehaghaji AA. Redistribution of fibres in the structure of hollow ring spun yarn. IJE Transactions B: Applications. 2009;**22**:197-204

[27] Hua T, Wong NS, Tang WM. Study on properties of elastic core-spun yarns containing a mix of spandex and pet/ptt bi-component filament as core. Textile Research Journal. 2018;**88**:1065-1076. DOI: 10.1177/0040517517693982

[28] Telli A, Daşan Y, Babaarslan O, Karaduman S. Usage of core and dual-core yarns containing tungsten for electromagnetic shielding. Advance Research in Textile Engineering. 2017;**2**:1-7

[29] El-Tantawy S, Sabry M, Bakry M. The effect of different weft yarn production technique on the pilling property of jeans fabrics. International Design Journal. 2017;**7**:161-169

[30] Bedez UT. Analysis of mechanical and dimensional properties of the denim fabrics produced with double-core and core-spun weft yarns with different weft densities. Journal of the Textile Institute. 2018;**109**:1-7. DOI: 10.1080/00405000.2018.1470451

[31] Musaddaq A, Amal B, Jakub W, Antonin H. Mechanism of liquid water transport in fabrics: A review. Fibers and Textiles. 2017;**4**:58-65

[32] Babaarslan O, Sarioğlu E, Çelik Hİ, Avci MA. Denim fabrics woven with dual core-spun yarns. In: Kumar Singh M, editor. Engineered Fabrics. London: InTech; 2018. pp. 19-39. DOI: 10.5772/intechopen.80286

[33] Fangueiro R, Filgueiras A, Soutinho F. Wicking behavior and drying capability of functional knitted fabrics. Textile Research Journal. 2010;**80**:1522-1530. DOI: 10.1177/0040517510361796

[34] Sarıçam C. Absorption, wicking and drying characteristics of compression garments. Journal of Engineered Fibers and Fabrics. 2015;**10**:146-154

[35] Ramesh Babu V, Ramakrishnan G, Subramanian VS, Lakshmi K. Analysis of fabrics structure on the character of wicking. Journal of Engineered Fibers and Fabrics. 2012;**7**:28-33

[36] Nassar K, Abou-Taleb EM. Effect of selected fabric construction elements on wicking rates of pet fabrics. Journal of Textile Science & Engineering. 2014;**4**:158. DOI: 10.4172/2165-8064.1000158

[37] Soloman C, Breckon T. Fundamentals of Digital Image Processing: A Practical Approach with Examples in Matlab. 1st ed. Chichester: Wiley; 2011. 344 p. DOI: 10.1002/9780470689776

[38] Ancutiene K, Koldinska M, Havelka A. Investigation of tensile resilience properties of strecth denim fabrics. Indian Journal of Fibre & Textile Research. 2017;**42**:175-182

[39] Ozdil N. Stretch and bagging properties of denim fabrics containing different rates of elastane. Fibres & Textiles in Eastern Europe. 2008;**16**:63-67

Chapter 3

Acoustic Insulation Behavior of Composite Nonwoven

Dilan Canan Çelikel and Osman Babaarslan

Additional information is available at the end of the chapter

http://dx.doi.org/10.5772/intechopen.80463

Abstract

Multilayer or multicomponent composite nonwoven structures provide great advantages for many technical applications. Spunbond-meltblown-spunbond (SMS) type multilayer nonwovens have significant commercial success in terms of end-use versatility. SMS type composite nonwovens, which can be produced with both continuous and discontinuous production technologies, are evolving day by day. Bulky, fibrous, porous nonwoven structures are widely used as sound absorbers for a variety of applications for instant building and automotive insulations, machine insulations, etc. The fibers interlocking in nonwovens are the frictional elements and provide resistance to acoustic wave motion. As many researchers reported, the most effective factors on sound absorption properties of fibrous materials are fiber diameter, airflow resistance, material thickness, tortuosity, porosity, and fiber surface area. In this research chapter, the sound absorption performance of SMS type composite nonwovens in relation to air permeability and pore sizes has been determined. The results show that SMS type nonwovens perform sound insulation at high frequencies. Spunmelt nonwovens with the advantages of short production line will create various alternatives with varieties of layering and compete with commercially used other sound absorbers.

Keywords: nonwovens, spunbond, meltblown, bicomponent fibers, sound absorption, permeability

1. Introduction

Today, textile materials are widely used in various technical applications including, but not limiting to transportation, agriculture, medical, filtration, and insulation. With the development of new materials, process technologies, combination of different processes, and innovative products have become the fastest growing area of the textile industry in recent years.

Combinations of different nonwovens, fiber combinations, and integration of process technologies are an increasingly beneficial option for new product developments.

The definition of nonwovens by EDANA (The European Disposables and Nonwovens Associations) and INDA (The North American Association of the Nonwoven Fabrics Industry), two leader associations in nonwovens market, "A nonwoven is a sheet of fibers, continuous filaments, or chopped yarns of any nature or origin, that have been formed into a web by any means, and bonded together by any means, with the exception of weaving or knitting." Nonwovens are engineered fabrics that can form the products that are disposable, for single or short-term use or durable, with a long life depending on the requirements and the intended product life [6, 7].

Multilayer or multicomponent nonwovens, also called as composite nonwovens, are the structures of various combinations of materials and processes providing great advantages. There are an increasing number of combinations of spunbonded (S) and meltblown (M) processes as SM, SMS, SMMS, SSMMMMSS, etc., where weaker meltblown fabrics are sandwiched between the stronger spunbonded fabrics. SMS (Spunbond + Meltblown + Spunbond) type multilayer nonwovens, has commercially success, are the best known products. These materials are produced continuously in a single line as well as a discontinuous line is also available.

Nonwovens with their bulky, fibrous, and porous structures, one of the most common textile materials, have an important role on sound absorption within the automotive, construction and a variety of industrial uses. Because of the porosity of the structure and the fibers interlocking in nonwovens are the frictional elements that provide resistance to acoustic wave motion. When sound enters into fibrous materials, its amplitude is decreased by friction as the waves try to move through the tortuous passages. Thus, the acoustic energy is converted into heat resulted with sound absorption [1].

In multilayer nonwovens, in accordance with the layers' structural parameters, different fiber intersections and fiber orientations occur. Pore connection and distribution become an important factor in determining acoustic properties because of the flow of sound wave through the material has been affected. Additionally as the number of layers increases or with different layer combinations in multilayer nonwovens, tortuosity and pore geometry will vary and different sound absorptions will be provided.

Many researchers studied and reported sound absorption characteristics of nonwovens/multilayer nonwovens in the literature. Ulcay et al. investigated sound absorption properties of spunbonded nonwovens produced from fibrillated islands in the sea bicomponent filaments with the various numbers of islands 1, 7, 19, 37, and 108. The results, as the effect of the number of islands on acoustical absorptive behavior, showed that spunbonded webs with 108 islands were better acoustic absorbers. Spunbonded nonwovens with the island in the sea bicomponent fibers were also compared with some high loft nonwovens; it has been reported that multilayer nonwovens with 108 islands have better sound absorbing performance [2].

Liu et al. studied the acoustic characteristics of dual-layered nonwovens by analyzing experimentally and theoretically. In experimental analysis, it was defined the sound absorption coefficients of 20 dual-layered nonwoven fabrics with four types of meltblown polypropylene nonwovens and five types of hydroentangled e-glass fiber nonwovens at low frequency ranges. In theoretical analysis, the effect of thickness and porosity of top and bottom layer on sound absorption coefficient was detailed using numerical simulation method. Experimental results indicated that the measured and the calculated data have very similar trend with the change of thickness, porosity, and the sound frequency [3].

Sound absorption properties of some bilayered needle-punched nonwoven composites at low frequencies have been investigated by Kucuk and Korkmaz. Results showed that macrofibrous layer of polyester fibers backed with 70% wool and 30% bicomponent polyester fibers has the best sound absorption properties at all frequency ranges [4].

Factors influencing acoustic performance of sound absorptive materials have been researched by Seddeq. As a result, he reported that the fiber linear density, air permeability, thickness, compression, porosity, and the position of the material are the major factors effecting acoustic properties of needle-punched nonwovens [5].

2. Multilayer nonwoven structures for acoustic insulation

Nonwoven industry had growth substantially for decades prior to the global recession between 2008 and 2010. The worldwide production of nonwovens was primarily based in Europe, North America, and Japan until the last decade. Now, nonwovens are produced on thousands of lines around the world. Asia is now the dominant nonwoven producing region, accounting for 42% of the world's production in 2014 [8]. Nonwoven production by region is shown in **Figure 1**.

The production of nonwovens is carried out in three stages as seen in **Figure 2**. Web formation is the major determinant of the characteristic of final product. The choice of methods for forming webs is determined mainly by fiber type and fiber length. The methods for the web formation from staple fibers were based on the drylaid and wetlaid processes, as well as in spunmelt processes, polymer chips are converted into webs by filament laying.

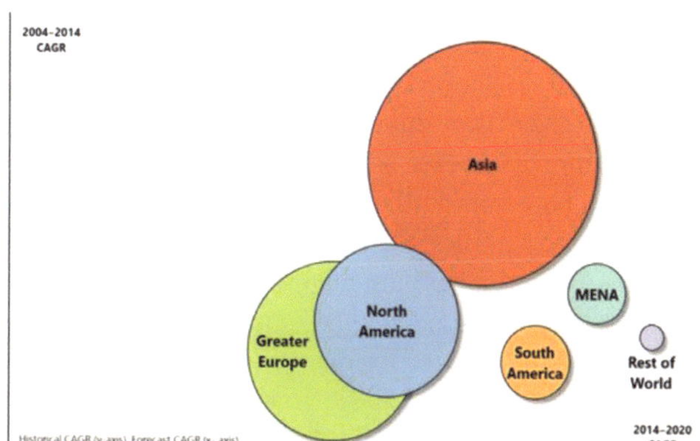

Figure 1. Nonwovens production by region (bubble size in tones of production) [8].

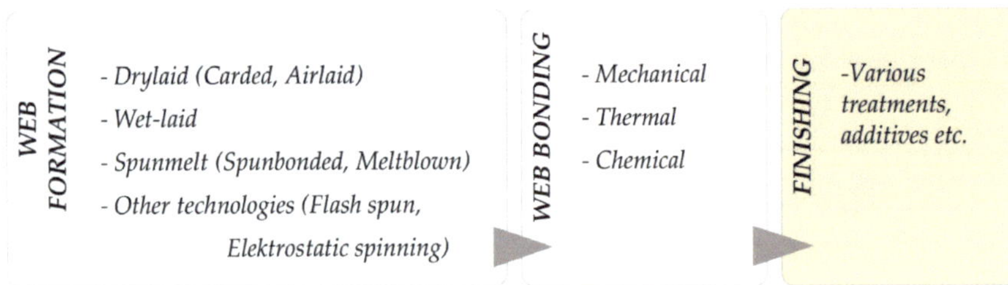

Figure 2. Nonwovens production stages.

Multilayer or composite nonwovens are produced by a modern and innovative industry that have numerous applications including, but not limiting to, hygiene, medical, filtration, insulation, automotive, agriculture, home furnishing, and packaging. Hygiene is the basic usage of multilayer nonwovens used in numerous products including baby care, feminine care, and adult products. The automotive industry also represents a significant market for application. Also breathable composite nonwovens are available for agriculture market. These materials offer engineering solutions by creating multifunctional products as well as economic solutions [14].

Sound absorbing materials are used in almost areas of noise control engineering to reduce sound pressure levels. They are used in a variety of locations—close to sources of noise, in various paths, and sometimes close to receivers. To use them effectively, it is necessary to:

- Identify the important physical attributes and parameters that cause a material to absorb sound.
- Provide a description of the acoustical performance of sound absorbers used to perform specific noise control functions.
- Develop experimental techniques to measure the acoustical parameters necessary to measure the acoustical parameters of sound absorbing materials and the acoustical performance of sound absorbers.
- Introduction of sound absorbing materials in noise control enclosures, covers, and wrappings to reduce reverberant build up and hence increase insertion loss.

Synthetic fibrous materials made from minerals and polymers are used mostly for sound absorption and thermal isolation. However, since they are made from high-temperature extrusion and industrial processes based on synthetic chemicals, often from petrochemical sources, their carbon footprints are quite significant. Although polyurethane and melamine foams are probably the cellular porous sound-absorbing materials currently most in use, other types of foams have been designed for environments where heat or corrosion resistance is required. Perforated panel absorbers have been used for many years in noise control usually to confine porous absorbing materials. When spaced away from a solid backing, a perforated panel is effectively made up of a large number of individual Helmholtz resonators, each consisting of a neck, comprised of the perforated panel and a shared air volume formed by the total volume of air enclosed by the panel and its backing. When the sound waves penetrate the perforated panel, the friction between the moving molecules of air and the internal surface of the perforations dissipates the acoustical energy into heat. The perforations are usually holes or slots, and as with a single resonator, porous material is usually included in the airspace to introduce damping into the system [30].

In this research chapter, it has been examined that the sound absorption characteristics of SMS type composite nonwovens. SMS structure is a spunmelt structure where the middle layer is the meltblown, sandwiched between the two top and bottom spunbonded layers.

Spunbonded and meltblown methods are both melt spinning method basically, with the shortest textile production line from polymer chips to a web. In the spunbonded method, continuous filaments are extruded directly from thermoplastic polymer chips. The formation of a web of continuous filaments deposited on the conveyor belt is assisted by air suction. Some residual temperature creates a weak bonding effect on the filaments but this is not considered as bonding. The web is then bonded directly by various means, normally thermal bonding. The web obtained is anisotropic. As thickness ranges from 0.2 to 1.5 mm, basis weights from 10 to 200 gsm. Filament thicknesses are between 10 and 80 μm [6, 13, 14].

Meltblown method is similar to spunbond. The hot, molten, low viscosity polymer is forced through nozzles to form a stream of polymer. At the nozzle tip, the filaments are picked up by hot, high velocity air streams that stretch the filaments by drag forces into very fine diameters. The filaments gradually cool as they travel across to the collector, a conveyor band or drum. The use of suction at the collector assists in web formation. The main typical characteristics for meltblown nonwovens are weak tensile properties, porous and capillary structure, isotropic formation, large surface area, etc. As the basis weight of the meltblown webs varies between 10 and 350 gsm, the fineness of the fibers ranges from 0.5 to 30 μm [9–11].

Spunbonded structures have a number of advantages as fabric's durability and lower cost in comparison to other nonwovens and woven and knitted fabrics. Meltblown nonwovens are made from microfibers that are much finer than in the spunbonded process. The fibers' fineness makes fabrics that are softer but much weaker than spunbonded materials. Due to the larger volume of fiber per unit weight, meltblown materials have improved fiber distribution and are important to a broad range of functional applications. For example, meltblown fabrics have good barrier properties and high insulating values and thus are used in filtration, barrier materials for medical and disposable apparel and apparel insulation. So together, the combination of spunbonded and meltblown structures can create a strong product which can also offer functional applications. Spunbonded layers act as protective layers for meltblown layers [10–12].

As the bonding method of spunmelt nonwovens, thermal bonding is usually available. Thermal bonding is the process to heat the web where heat is treated with hot rollers, hot air or sound waves. It can be carried out by means of heated calender rollers even after web formation in spunmelt systems. In these methods, fusing fibers act as thermal binders. Important process parameters affecting the web properties are roller temperature, roller speed or contact time and pressure applied to web. Additionally, roller pattern (flat, pointed, etc.) controls the fabric strength, drape, stiffness, and softness. Heating temperature of rollers should be suitable for melting point of the polymer consisting of web [9–11].

One of the applications in thermal bonding in spunbonded process is producing web consisting of bicomponent fibers. Bicomponent fibers contain two different polymers extruded together from the same spinneret to compose a single fiber cross section. The properties and applications of bicomponent fibers depend on both the properties and distribution of the polymers in the cross-sectional area. Accordingly, typical configurations are side by side, core/sheath, island in the sea, sliced, pie slice, etc. The most commonly used in nonwovens and well-known binding bicomponent fibers is sheath/core type. When a bicomponent nonwoven web is heated sufficiently to melt the sheath, polymer melts and flows to the nearest adjacent fiber and binds the structure. It is recommended that the melting temperature difference between the components should be at least 40°C for proper bonding. Lower bonding temperature is provided by bicomponent fibers than in a typical thermal bonding application. Additionally, with this method, some structural parameters of a nonwoven fabric, such as fabric density, fiber diameter, tortuosity, porosity, etc., will be affected [15, 20].

3. Materials and methods

3.1. Materials

In this research, all spunbonded and meltblown nonwoven fabrics were supplied by Mogul Nonwoven Company in Gaziantep/Turkey. Raw material of all layers was polyester with the advantages of availability, flexibility, and commercially success. Spunbonded layers, flat bonded thermally, having a basis weight of 40 gsm, were produced from homocomponent and bicomponent fibers in the diameter of 20–24 μm. Seven different meltblown layers had a basis weight of 50–200 gsm with homocomponent round fibers at the diameter of 5–8 μm. Meltblown nonwovens were bonded thermally at same conditions to form nonstiffer middle layer of multilayer structures. SMS compositions of two different spunbonded layers and seven different meltblown layers were prepared manually, resulting in 14 multilayer nonwoven structures. Layers were arranged loosely, adjacent to each other. Fabric design of multilayer structures has been illustrated in **Figure 3**.

Description of multilayer nonwoven samples is shown in **Table 1**. The thickness of the samples ranged from 1.28 to 1.79 mm. From **Table 1**, the samples were coded as HC and BC according to the change of fiber type in the spunbond layers; sample codes of 1, 2, 3, 4, 5, 6, and 7 defined the changes in basis weight of the meltblown layers. For instance, HC1 designates an SMS type three-layered nonwoven in which the outer spunbonded layers with homocomponent round fibers and a meltblown layer have a basis weight of 50 gsm; BC7 means spunbonded layers with bicomponent fibers and meltblown layer having a basis weight of 200 gsm.

In this research, bicomponent fibers, round core/sheath type, in the spunbonded layers have a polyester core with an outer sheath of copolyester. The composition of polymers in core/sheath type is 90% polyester (PET) core with 230–250°C melting point and 10% copolyester (Co-PET) sheath with 110–140°C melting point.

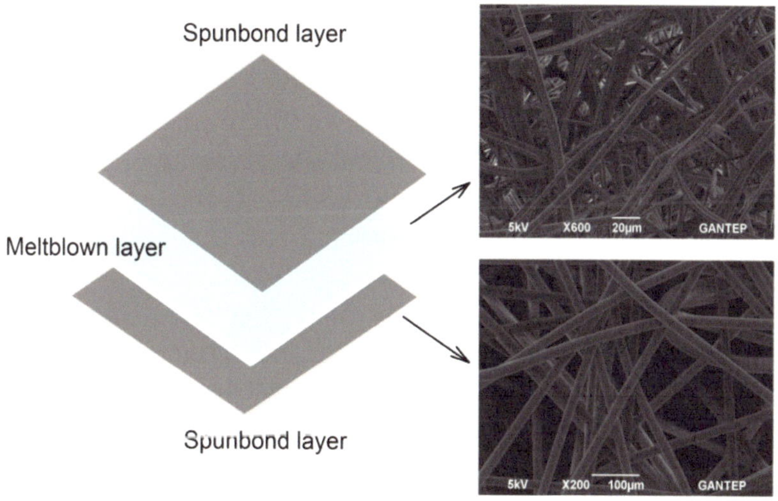

Figure 3. Schematic of multilayer structure and SEM micrographs of spunbonded layers (homocomponent fibers) and meltblown layer (in basis weight of 125 gsm).

Sample ID	Type of layer	Fiber type	Fiber content	Fiber cross section	Basis weight (gsm)	Thickness (mm)
HC	Spunbonded layers	Homocomponent	PET	Round	40	0.37 ± 0.07
BC		Bicomponent	PET/Co-PET	Bico-round/sheath-core type		0.35 ± 0.05
1	Meltblown layer	Homocomponent	PET	Round	50	0.59 ± 0.09
2					75	0.60 ± 0.08
3					100	0.62 ± 0.06
4					125	0.67 ± 0.07
5					150	0.84 ± 0.05
6					175	0.93 ± 0.06
7					200	1.1 ± 0.04

Table 1. Sample description and specifications of nonwoven layers.

3.2. Methods

All measurements were carried out at standard temperature (20 ± 2°C) and relative humidity (65 ± 2%). The thickness of the 10 different samples from each material was measured using a standard measuring device according to NWSP 120.6.R0 (15) [16].

Air permeability of multilayer nonwovens was obtained by using an SDL Atlas digital air permeability tester (SDL-Atlas Inc., USA). The test were conducted according to NWSP 070.1.R0

(15) [17]. The measurements were done on five different samples from each material by applying 200 Pa pressure through a 20 cm^2 test area. The reported results are the averages of the five measurements.

Sound absorption coefficients of multilayer nonwovens were measured according to ISO 10534-2 [18]. Nonwoven samples were cut into 100 and 29 mm diameters for the measurement of large and small tubes. Sound absorption coefficients of three samples (two replications from each material) were obtained by using a Brüel and Kjær impedance tube kit (**Figure 4**).

The capillary flow porometer (Porous Materials Inc., USA) has been successfully used to evaluate pore structures of multilayer nonwovens. Determination of porosity of samples according to ISO 15901-1 standard, 5 samples were prepared at 0.03 cm and determined by taking the average of the measurement values [19].

In statistical analysis, Design Expert Analysis of Variance (ANOVA) software (Stat-Ease, Inc., USA) was achieved. The effect of independent parameters, basis weight (A) and fiber type (B) on the dependent parameters of air permeability, mean pore diameter, and sound absorption has been examined with the analysis of variance at significance level of p value less than 0.05.

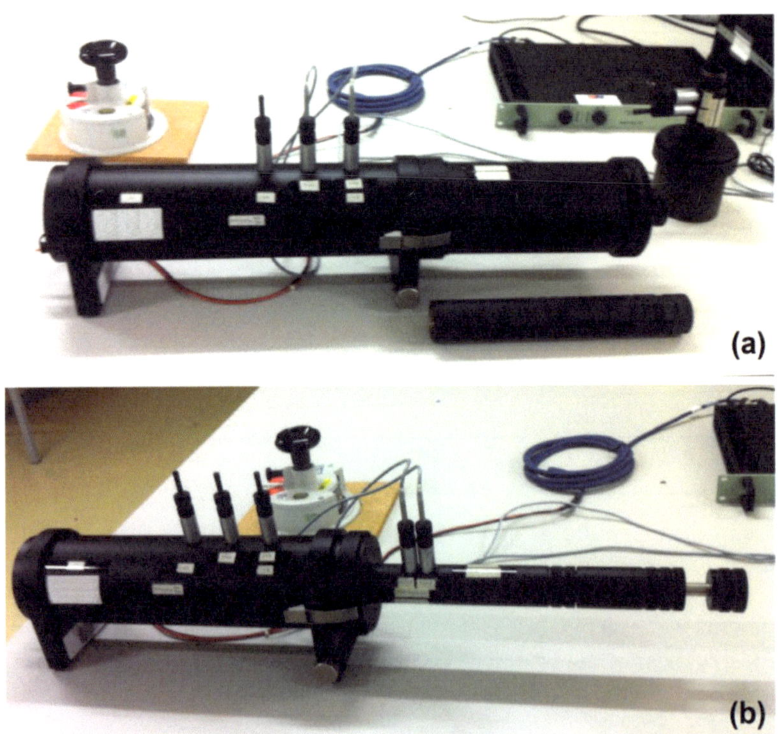

Figure 4. Impedance tubes for the two microphone transfer function methods: (a) large tube for 0.5–6.4 kHz and (b) small tube for 0.5–1.6 kHz.

4. Experimental results and discussion

4.1. Air permeability

Air permeabilities of nonwoven samples are presented in **Figure 5**, respectively, for increasing basis weights of multilayer nonwovens. As seen in **Figure 5**, the air permeabilities of the nonwoven samples with bicomponent fibers are lower than the air permeabilities of the nonwoven samples with homocomponent fibers. For each range of basis weight, BC samples are more resistant to air flow than the HC samples.

Air permeability is expressed as the ratio of air flow between the two surfaces of the fabric. The speed of the air flow passing vertically from a given area is measured by the pressure difference within the measuring area of the fabric. The degree of air permeability is one of the major affecting parameters of thermal and acoustic insulation capabilities of nonwoven fabrics. Higher air permeability results in higher sound absorption [21–23].

Co-PET polymer with low melting temperature in nonwoven samples containing bicomponent fibers melts during the thermal bonding and provides the binding by spreading to the fibers around the web. This attribute limits the cross sectional and connection between fibers, and when considering that it affects the fiber roughness, the decrease in pore diameters is determined by the pore size measurements. When the relationship between air permeability and pore structure is evaluated, it is thought that this will increase air flow resistance and create a decrease in air permeability values. This indicates that bicomponent structures restricted the size of air passages, so that air permeability decreased. At higher basis weights of the fabrics, the increase in the number of fibers creates more spaces and a longer tortuous path through which the air must flow. Thus fabric structure becomes more resistant to air flow resulted with lower air permeabilities.

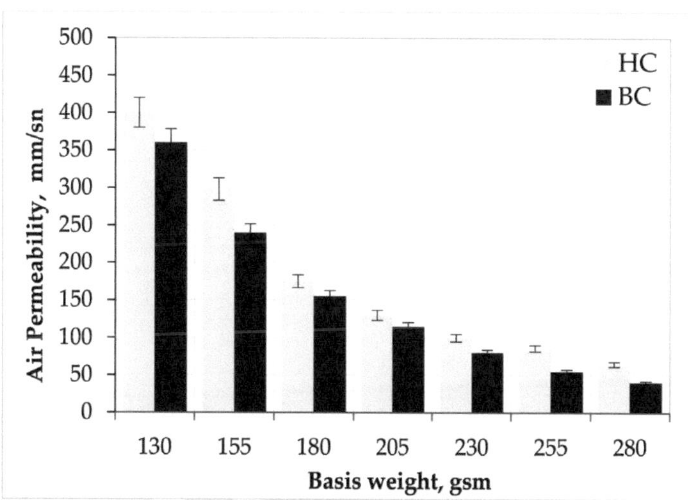

Figure 5. Air permeability of nonwoven samples.

In the statistical data analysis, the effect of independent parameters, basis weight (A) and fiber type (B), on the dependent parameter, air permeability, has been examined with the analysis of variance at significance level of p value less than 0.05. The model summary statistics and ANOVA results for the data obtained in the study are shown in **Table 2**.

As presented in **Table 2**, R-Squared (R^2) equals 0.9938, and predicted R-Squared (R_{pre}^2) equals 0.9865 for the model. It means that dependent parameters have been affected by independent parameters 99.38%, and this model predicts air permeability successfully at very high proportion of 98.65%. In the ANOVA results of BC and HC samples, both A and B are significant model terms. Contribution to model of significant terms according to F values, it has been determined that A-basis weight is more significant factor with higher F value for air permeability than B-fiber type. It can be specified that fiber type (bicomponent and homocomponent), is less effective parameter than basis weight to control air permeability of multilayer nonwovens statistically. Regression equation for air permeability of BC and HC samples obtained from the model is presented below in Eq. (1) according to codded factors. The high levels of the factors are coded as +1, and the low levels of the factors are coded as −1.

$$\text{Air permeability} = +121.00 - 124.34^{*} A - 14.93^{*} B + 97.23^{*} A^2 - 40.50^{*} A^3 \quad (1)$$

4.2. Mean pore diameter

The porosity of the fabrics is a complex feature characterized by parameters such as pore diameter, pore distribution, pore volume, while the porosity of the fabrics is associated with the total fabric volume area of the empty volume. Fabric porosity directly affects permeability properties, and the shape, layout, and size distribution of the media spaces are important considering the flow from porous structure [22].

In **Figure 6**, it is presented the mean pore diameter of nonwoven samples with the change of basis weight. As seen in **Figure 6**, nonwoven samples with bicomponent fibers had lower mean pore diameters than nonwoven samples with homocomponent fibers.

Source	Sum of squares	Degree of freedom (df)	Mean square	F	Significance
Model	1.748E + 005	4	43703.18	362.97	<0.0001
A-Basis weight	13084.75	1	13084.75	108.67	<0.0001
B-Fiber type	3120.07	1	3120.07	25.91	0.0007
Factors within group	20580.48	2			
Residual	1083.64	9	120.40		
Cor total	1.759E + 005	13			
Model	Std. deviation	10.97	R-Squared		0.9938
	C.V.%	6.68	Adjusted R-Squared		0.9911
	PRESS	2380.21	Predicted R-Squared		0.9865

Table 2. ANOVA for air permeability of samples.

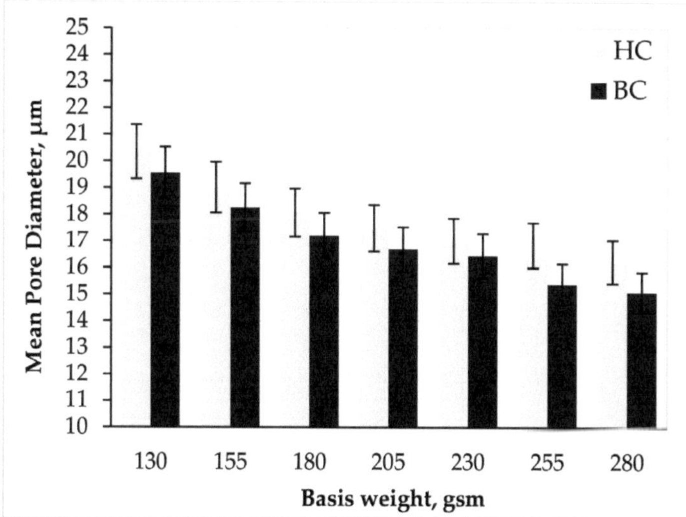

Figure 6. Mean pore diameter of nonwoven samples.

Porosity, thickness, and fiber diameter are the factors that affect tortuosity of the structure [24]. Co-PET polymer with low melting point in the bicomponent nonwoven samples melted earlier and smeared the adjacent fibers during the bonding. The reason may be the variation of intersection of fibers, roughness, and tortuosity resulted with the change of the pore structure as smaller pore diameters for bicomponent nonwovens.

The statistical analysis of mean pore diameter of BC and HC samples exhibited in **Table 3** indicates the significant effect of fiber type with higher F values. Mean pore diameters have been affected by basis weight and fiber type 98.97%, and the model predicts the actual values of air permeability 97.60%. It can be specified that fiber type (bicomponent and homocomponent)

Source	Sum of squares	Degree of freedom (df)	Mean square	F	Significance
Model	29.47	4	7.37	216.22	<0.0001
A-Basis weight	1.92	1	1.92	56.46	<0.0001
B-Fiber type	2.91	1	2.91	85.34	<0.0001
Factors within group	1.28	2			
Residual	0.31	9	0.034		
Cor total	29.78	13			
Model	Std. deviation	0.18	R-Squared		0.9897
	C.V.%	1.06	Adjusted R-Squared		0.9851
	PRESS	0.71	Predicted R-Squared		0.9760

Table 3. ANOVA for mean pore diameter of samples.

is more effective parameter than basis weight to control mean pore diameters of multilayer nonwovens statistically.

Regression equation for air permeability of BC and HC samples obtained from the model is presented below in Eq. (2) according to codded factors. The high levels of the factors are coded as +1, and the low levels of the factors are coded as −1.

$$\text{Mean pore diameter} = +17.09 - 1.51^{*}\text{A} + 0.46^{*}\text{B} + 0.70^{*}\text{A}^{2} - 0.65^{*}\text{A}^{3} \quad (2)$$

4.3. Sound absorption

The performance of sound absorbing materials is generally explained by the sound absorption coefficient (α). It is defined as the ratio of acoustic energy that is trapped in the material by the material and ranges between 0 and 1. "α = 0" means 0% sound absorption so the reflection of all the sound waves, and "α = 1" means 100% sound absorption of all the sound waves.

The sound absorption results of nonwoven samples are observed in **Figure 7**. It is certain that as many researchers reported, the increase in basis weight influences the sound absorption positively. So also in this research, the higher sound absorption coefficients were proved for the higher weights. But it should be noted that BC samples have better sound insulation for each range of fabric weight. More effective sound absorption with bicomponent fibers is obvious.

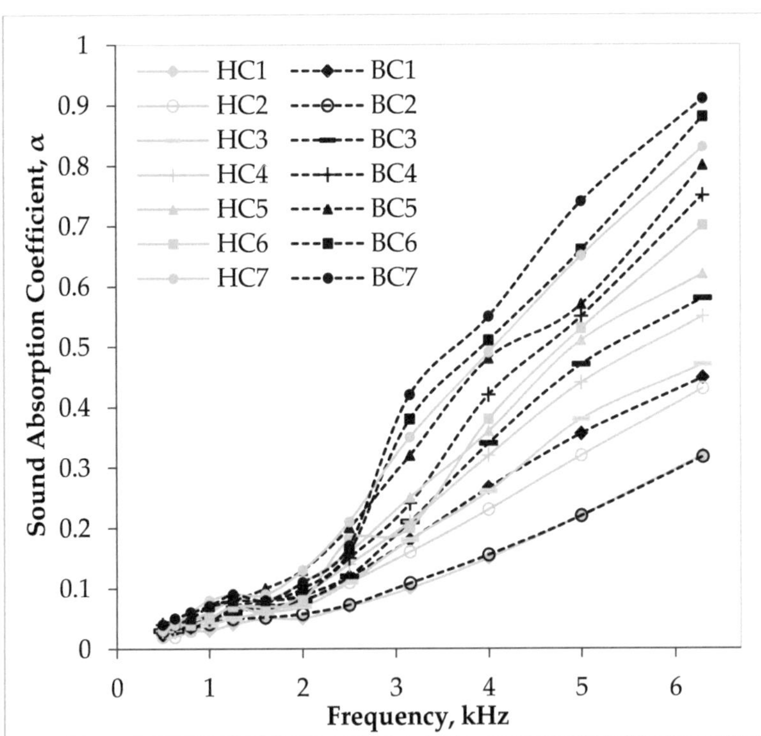

Figure 7. Sound absorption coefficients of nonwoven samples.

The effectiveness of a material in sound absorption depends mainly on the frequency of the sound wave subjected to the material, basis weight, air permeability, fiber geometry, and fiber arrangement [25–27]. Sound absorption occurs due to the impact of sound waves on material, friction losses while moving in the pores and channels of the structure, and the decrease in sound energy. As a result of increasing basis weight, fiber density, and porosity of random fibers, the sound wave will contact more fibers, and friction losses will increase [23]. As a result, the sound energy will be reduced, and higher sound absorption coefficients will be obtained. The results obtained from this research are evidence of this situation.

In bicomponent fibers with core/sheath type round cross section, melting Co-PET smeared to adjacent fibers to bind the nonwoven structure. It can be concluded that melting part of bicomponent fibers affects the cross-section area and fiber surface roughness, resulted with the variation of tortuous passages performed as higher sound absorption. As the result of restricted flow of sound waves, the sound absorption coefficients became higher [28, 29].

The statistical analysis of sound absorption of BC and HC samples presented in **Table 4** indicates the significant effect of fiber type with higher F values. Sound absorption has been affected by basis weight and fiber type 98.98%. It can be specified that fiber type (bicomponent and homocomponent) is more effective parameter than basis weight to control sound absorption of multilayer nonwovens statistically.

Regression equation for air permeability of BC and HC samples obtained from the model is presented below in Eq. (3) according to codded factors. The high levels of the factors are coded as +1, and the low levels of the factors are coded as −1.

$$\text{Sound absorption} = +0.65 + 0.25{*}\,A + 0.10{*}\,B + 0.01\,2{*}\,AB - 0.25{*}\,A^2 - 0.30{*}\,A^2B + 0.24{*}\,A^3 + 0.41{*}\,A^3B + 0.23{*}\,A^4 + 0.25{*}\,A^4B - 0.25{*}\,A^5 - 0.44{*}\,A^5B \quad (3)$$

Source	Sum of squares	Degree of freedom (df)	Mean square	F	Significance
Model	0.53	11	0.048	900.08	0.0011
A-Basis weight	0.012	1	0.012	230.51	0.0043
B-Fiber type	0.036	1	0.036	685.34	0.0015
Factors within group	0.005	9			
Residual	1.061E-004	2	5.303E-005		
Cor total	0.53	13			
Model	Std. deviation	7.282E-003	R-Squared		0.9998
	C.V. %	1.18	Adjusted R-Squared		0.9987
	PRESS	0.20	Predicted R-Squared		0.6143

Table 4. ANOVA for sound absorption of samples.

	A: Basis weight	B: Fiber type	Air permeability	Mean pore diameter	Sound absorption
A: Basis weight	1.000	0.000	−0.927	−0.921	0.911
B: Fiber type	0.000	1.000	−0.133	0.313	0.284
Air permeability	−0.927	−0.133	1.000	0.882	−0.869
Mean pore diameter	−0.921	0.313	0.882	1.000	−0.741
Sound absorption	0.911	0.284	−0.869	−0.741	1.000

Table 5. Correlation between variables and responses.

4.4. Correlations

The correlations between the independent parameters, basis weight (A) and fiber type (B), and the dependent parameters, air permeability (C), mean pore diameter (D), and sound absorption (E), has been examined and the results for the data obtained in the study are shown in **Table 5**. The statistical results have showed that air permeability, mean pore diameter, and sound absorption have significant correlation. Correlations between each of variables proved that sound absorption has an inverse relation with air permeability and pore sizes.

5. Conclusion

In this research, acoustic insulation behavior of SMS type composite nonwovens has been investigated. The results show that sound absorption has been affected by fiber type of homocomponent or bicomponent, and more effective sound absorption with bicomponent fibers is obvious. Higher sound absorption coefficients were provided with multilayer nonwoven samples containing bicomponent fibers.

The reason for these results may be because the different porosity, tortuosity, and roughness of bicomponent and homocomponent structures. Higher value of tortuosity would therefore indicate longer, more complicated, and sinuous path, thus resulting in greater resistance to sound wave flow. Tortuosity also directly influences propagation of acoustic waves and absorbance efficiency in fibrous porous media. It has also been said that the degree of tortuosity determines the high frequency behavior of sound absorbing porous materials.

Additionally, sound absorbent materials must be porous in order to allow sound waves to enter, spread, and decrease sound energy through friction. However, closed pores in the structure have little effect on the absorption of sound, while open pores directly affect the sound insulation properties of the material as they allow sound waves to penetrate into the material [21, 30].

It can be stated that this effect increases with the contribution of bicomponent fibers in the formation of nonwoven surface with filament laying methods, which constitute the basic character of the process of random fiber orientation and intersection [31–34]. As a result, while

http://dx.doi.org/10.5772/intechopen.80463

the pore diameters and air permeability values of the samples containing bicomponent fibers decrease, friction of the sound wave by changing direction with the fiber surfaces can be said to cause loss of acoustic energy and increase of sound absorption coefficients.

When sound absorption and effecting factors are evaluated on nonwovens, the basis weight, thickness, fiber fineness, air flow resistance, pore structure, and tortuosity can be listed. Sound absorption performances of surfaces with high resistance to air flow up to a certain point will also be high. However, according to the studies in the literature, this situation may vary at various levels depending on the conditions of use and the frequency of the sound wave. The air permeability is mainly effective in determining the sound absorption, and the structural parameters that control the air permeability will also affect the sound absorption. The results obtained in this study show that the air permeability values of samples with smaller pore diameters are also lower, supporting the high sound absorption of these samples. In addition, it has been determined that this effect has become more pronounced with the increase of basis weight. As fabric basis weights ranged from 130 to 280 gsm for each sample group, as the increase in the number of fiber per unit area, the higher sound absorption coefficients were obtained. Increasing intersection of fibers in heavier nonwovens creates a tortuous path for sound wave to flow caused to acoustic energy loss and sound absorption.

Additionally, all samples had low sound absorption coefficient range between 0.0 and 0.3 up to the frequency of 3000 Hz.

At the high frequencies, as the wavelengths becomes smaller, the thinner fabrics control the sound absorption efficiently. Therefore, the thinner spunmelt nonwovens compared to needle-punched ones are good sound absorbers at high frequencies.

The results show that sound insulation at high frequencies can be improved by using spunmelt multilayer nonwovens. Spunmelt multilayer nonwovens offer opportunities to tailor fabrics to desired applications through variations in fiber type and basis weight.

Acknowledgements

This research has been supported by Scientific Research Project Department of Cukurova University. The authors would like to thank also Mogul Nonwoven Company/Gaziantep/Turkey for supplying nonwoven samples.

Author details

Dilan Canan Çelikel[1]* and Osman Babaarslan[2]

*Address all correspondence to: celikel@gantep.edu.tr

1 Gaziantep University, Technical Sciences Vocational School, Gaziantep, Turkey

2 Çukurova University, Department of Textile Engineering, Adana, Turkey

References

[1] Anantharamaiah N, Verenich S, Pourdeyhimi B. Durable nonwoven fabrics via fracturing bicomponent islands-in-the-sea filaments. Journal of Engineered Fibers and Fabrics. 2008;**3**(3):1-9

[2] Suvari F, Ulcay Y, Maze B, Pourdeyhimi B. Acoustical absorptive properties of spunbonded nonwovens made from islands-in-the-sea bicomponent filaments. The Journal of The Textile Institute. 2013;**104**:438-445

[3] Liu J, Liu X, Xu Y, Bao W. The Acoustic characteristics of dual-layered porous nonwovens: A theoretical and experimental analysis. The Journal of The Textile Institute. 2014;**105**:1076-1088

[4] Kucuk M, Korkmaz Y. Sound absorption properties of bilayered nonwoven composites. Fibers and Polymers. 2015;**16**(4):941-948

[5] Seddeq HS. Factors influencing acoustic performance of sound absorptive materials. Australian Journal of Basic and Applied Sciences. 2009;**3**:4610-4617

[6] Russel SJ. Handbook of Nonwovens. Manchester, England: Textile Institute, Woodhead Publishing; 2007. Available from: https://books.google.com.tr./ [Accessed: March 10, 2018]

[7] Definition of Nonwovens [Internet]. Available from: https://www.edana.org/discover-nonwovens/what-are-nonwovens [Accessed: March 10, 2018]

[8] Worldwide Outlook for the Nonwovens Industry. INDA/EDANA [Internet]. 2015. Available from: http://www.metissue.com/article/nonwovens-industry-outlook/ [Accessed: April 10, 2018]

[9] Albrecht W, Fuchs H, Kittelmann W. Nonwoven Fabrics: Raw Materials, Manufacture, Applications, Characteristics, Testing Processes. John Wiley & Sons Press; 2006. Available from: https://books.google.com.tr./ [Accessed: March 20, 2018]

[10] Karthik T, Prabha Karan C, Rathinamoorthy R. Nonwovens: Process, Structure, Properties and Applications. CRC Press; 2017. Available from: https://books.google.com.tr./ [Accessed: March 20, 2018]

[11] Butler I. The Spunbonded and Melt Blown Technology Handbook. INDA International Nonwovens Consulting, Inc; 1999. Available from: https://imisw.inda.org/CMDownload.aspx?...5add [Accessed: March 10, 2018]

[12] Dutton CK. Overview and analysis meltblown process and parameters. Journal of Textile and Apparel, Technology and Management. 2008;**6**(1):1-25

[13] Bhat GS. Extruded continuous filament nonwovens: Advances in scientific aspects. Journal of Applied Polymer Science. 2001;**83**(3):572-585

[14] Das D, Pourdeyhimi B. Composite Nonwoven Materials, Structure, Properties and Applications. Woodhead Publishing; 2014. Available from: https://books.google.com.tr./ [Accessed: March 30, 2018]

[15] Dasdemir M, Maze B, Ananthsaramaiah N, Pourdeyhimi B. Influence of polymer type, composition, and interface on the structural and mechanical properties of core/sheath type bicomponent nonwoven fibers. Journal of Materials Science. 2002;**47**:5955-5969

[16] NWSP 120.6.R0 (15): Nonwoven Thickness (EDANA)

[17] NWSP 070.1.R0 (15): Air Permeability of Nonwoven Materials

[18] ISO 10534-2 Determination of Sound Absorption Coefficient and Impedance in Impedance Tubes–Transfer Function Method

[19] ISO 15901-1 Evaluation of Pore Size Distribution and Porosity of Solid Materials by Mercury Porosimetry and Gas Adsorption

[20] Fedorova N, Pourdeyhimi B. High strength nylon micro- and nanofiber based nonwovens via spunbonding. Journal of Applied Polymer Science. 2007;**104**:3434-3442

[21] Marmaralı A, Ertekin G, Çay A. Çeşitli kumaş parametrelerinin yuvarlak örme sandviç kumaşlarin ses yutum özelliklerine etkisi. Pamukkale Üniversitesi Mühendislik Bilim Dergisi. 2014;**20**(7):281-286

[22] Okur A, Turan RB. Air permeability of fabrics. In: Textile and Engineer. TMMOB Tekstil Mühendisleri Odası Yayınları; 2008

[23] Tascan M, Lyon Gaffney K. Effect of glass-beads on sound insulation properties of nonwoven fabrics. Journal of Engineered Fibers and Fabrics. 2012;**7**(1):101-105

[24] Vallabh R, Banks-Lee P, Seyam AF. New approach for determining tortuosity in fibrous porous media. Journal of Engineered Fibers and Fabrics. 2010;**5**:7-15

[25] Lee YE, Joo CW. Sound absorption properties of thermally bonded nonwovens based on composing fibers and production parameters. Journal of Applied Polymer Science. 2004;**92**:2295-2302

[26] Midha VK, Chavhan MV. Nonwoven sound absorption materials. International Journal of Textile and Fashion Technology. 2012;**2**(2):45-55

[27] Midha VK, Dakiri A. Spun bonding technology and fabric properties: A review. Journal of Textile Engineering & Fashion Technology. 2017;**1**(4):1-9

[28] Rawal A. Structural analysis of pore size distribution of nonwovens. The Journal of The Textile Institute. 2010;**101**(4):350-359

[29] Castagnede B, Aknine A, Brouard B, Tarnow V. Efects of compression on the soundabsorption of fibrous materials. Journal of Applied Acoustics. 2000;**61**:173-182

[30] Arenas JP, Crocker MJ. Recent trends in porous sound-absorbing materials. Journal of Sound and Vibration. 2010

[31] Krucin I, Gliścin E, Michalak M, Ciechan D, Kazimierczak J, Bloda A. Sound-absorbing green composites based on cellulose ultra-short/ultra-fine fibers. Textile Research Journal. 2016:647-657

[32] Wenbin ZW, Nandikolla W, George B. Effect of bulk density on the acoustic performance of thermally bonded nonwovens. Journal of Engineered Fibers and Fabrics. 2015;**10**(3):39-45

[33] Coates M, Kierzkowski M. Acoustic textiles. Technical Textile International Conference. 2002

[34] Shahani F, Soltani P, Mohammad ZM. The analysis of acoustic characteristics and sound absorption coefficient of needle punched nonwoven fabrics. Journal of Engineered Fibers and Fabrics. 2014;**9**(2):84-92

Chapter 4

Lace Braiding Machines for Composite Preform Manufacture

David Branscomb, Yang Shen, Vladimir Quinones,
Royall Broughton and David Beale

Additional information is available at the end of the chapter

http://dx.doi.org/10.5772/intechopen.82256

Abstract

This paper is an evaluation of a modern lace braiding machine technology for suitability in the manufacture of textile composite material preforms. A brief history of bobbin lace and lace braiding machines is provided along with a discussion of the functionality of a Barmen lace braiding machine—the predecessor to the modern computerized lace braiding machine. It was found that the typical modern lace braiding machine lacked the robustness necessary to produce braided preforms using large, high-strength synthetic yarns such as carbon and aramid that are commonly found in advanced composite materials. Improvements are proposed to enable lace braiding machines to be developed for future applications.

Keywords: lace, jacquard, barmen, torchon, preform

1. Introduction

Lace braided fabrics embody intricate patterns consisting of precisely placed yarns into structures which from an advanced fiber placement standpoint appear suitable for pre-forms in composite manufacture. Lace is characterized by openwork or lattice architecture with regular and irregular patterns propagating the fabric. Much like trusses that offer high strength or stiffness-to-weight reinforcement through strategic reinforcement placement, lace formation technologies could be used to produce efficient textile preforms or composite space trusses directly. The intrinsic structural characteristics of lace such as openness and precision fiber placement could be utilized to eliminate expensive secondary operations such as drilling.

Machining of holes is required for fastening and joining of ancillary components while potentially increasing reinforcement efficiency and minimizing weight. Lace machines can produce either flat or seamless cylindrical structures.

To provide a thorough evaluation of lace braiding technology for composite material manufacture, a brief historical context of industrial lace manufacture along with details of the development of the present-day lace braiding machine is presented. The fundamental features of braided lace are detailed to form a foundation for establishing requisite engineering design fundamentals, and to evaluate the present-day lace braiding machine for immediate suitability to produce structural composite reinforcements.

We find little evidence to suggest that braided lace has been used significantly for structural composite reinforcements. Braided lace does not readily appear to be available in the formation of heavy or industrial fabrics which would imply suitability in advanced composite preforms commonly using 12k and greater carbon fiber tows. However, one manufacturer currently offers engineered lace patterns for a myriad of applications including soft composites, sporting goods, and advanced apparel [1]. Several sources describe the esoteric nature and the lack of readily available design information [2, 3]. Although, lace braiding machines have been used in medical applications and smart textiles [4]. In fact, one reference suggests that lace braiding has no application to composites. For example, The Handbook of Composite Reinforcements provides a list of braiding machines according to the structures formed and the application to composites [5]. Per Lee, the Jacquard braiding machine (a.k.a. lace braiding machine) is used to produce tubes and flat strips with complex lace patterns and does not list composites as being an application. Work evaluating the Jacquard mechanism used in commercially available machines and proposed improvements in the control scheme are presented by Yang [6]. Many articles exist on various details of pillow lace formation techniques, a handful of short articles are dedicated to certain manufacturing aspects of machine lace (mostly trade publications), and very little can be found outside of the patent literature regarding the mechanics of the machines. However, a comprehensive text on the history and development of various lace machines by Earnshaw is recommended to the reader for additional study of the broader topic [7].

2. History of lace, lace machine development, and modern lace machines

Originally lace was produced by hand as a highly skilled art form requiring years of experience due to its almost infinite design possibilities. Handmade lace can be traced to the fifteenth century in Italy, Spain, Germany, and the Netherlands [8]. **Figure 1** shows bobbin or pillow lace as an example. Although there are many forms of openwork lace such as crochet, knitting, and tatting, bobbin lace specifically is formed by braiding.

Various types of lace producing machines exist, such as Raschel and Leavers machines, but the operation mechanisms are fundamentally different from the lace braiding machines. John Heathcoat's bobbinet machine is arguably one of the original lace machines. The lucrative manufacture of lace led to numerous patents issued during the mid-1800s and ultimately to the mass manufacture of lace [10].

http://dx.doi.org/10.5772/intechopen.82256

Figure 1. Handmade bobbin lace or pillow lace [9].

The dexterous movements performed in bobbin or pillow lace formation have many similarities with the mechanical movements of the lace braiding machine. The fundamental movements are therefore similar. This is not surprising as the lace braiding machines of the nineteenth century were expressly designed to mimic the motion of bobbin lace makers' hands. One of the first patents issued for a bobbin lace braiding machine was issued in 1910 to Gustav Krenzler of Barmen, Germany [11]. Earlier in the same year, a patent was issued to Emil Krenzler for a single-thread lace-bobbin machine [12]. Lace braiding machines have several names including, Barmen, Torchon, and Jacquard lace braiding machines. These machines produce tubular fabrics which are then separated into two flat fabrics of the same design for efficiency. Small monofilament yarns are used to join the two "flat" fabrics which are subsequently removed.

3. Fundamentals of lace braiding

Lace braiding machines utilize rotating plates, analogous in functionality to horn gears of Maypole braiding machines, to control bobbin motion and produce desired designs. To the credit of machine lace and a testament of its versatility, it can be difficult to distinguish from its handmade counterparts [7]. Some limitations are imposed by mechanical aspects of the machine design; however, simple laces such as Torchon lace can be easily made with the lace braiding machine [13].

Braided lace is formed by basic stitches typically applied to bobbin pairs. In this case, a pair of bobbins may simply rotate clockwise or counterclockwise as a twisted pair as well as interchanging with an adjacent bobbin pair. Individual control of a single yarn or yarn pair enables lace designs to be complex with almost infinite possibilities. However, even the most complex designs are derived from two motions. These basic motions comprise various stitches and by combining simple motions, intricate patterns may be designed. In general, the design of lace is described by the stitches, i.e., the basic movements of bobbin pairs. Furthermore, by utilizing various materials and yarn tensions, other desired features such as textures and holes may be imparted to the lace.

4. Stitches of braided lace

During the formation of lace, the yarns form an X that is identified as either a cross or a twist depending on the direction. Twist is defined as a counter-clockwise motion where the right yarn of each pair is laid over the left yarn. The twist motion pairs stay together on the machine plate. Cross is a clockwise motion worked with two adjacent pairs and the inner pairs are crossed so that the left yarn of the inner pair is laid over the right yarn. The cross pairs are interchanged. In **Figure 2**, the first and fourth yarns remain stationary while the second and third yarns cross multiple times. Then the first and third yarn twists multiple times simultaneously with the second and fourth yarns.

4.1. Barmen lace

For various reasons details of the Barmen braiding machine have not been readily available, and known only by a select few. The complexity of these machines tended to require specialized operational expertise as well. Thus, expertise with these machines tended to be concentrated within the immediate geographic region of the machine origin. In the case of the Barmen lace braiding machine, the region was Barmen—an industrial city that later merged with Wuppertal, Germany. Publications, outside of textile trade literature and patent literature, related to the Barmen lace braiding machine are scarce. The descriptions found in the patent literature are inadequate for interested audiences outside those skilled in the art. In general, the lace braiding industry had many trade secrets where knowledge of pattern design, machinery, and operations was confined to an esoteric group of practitioners.

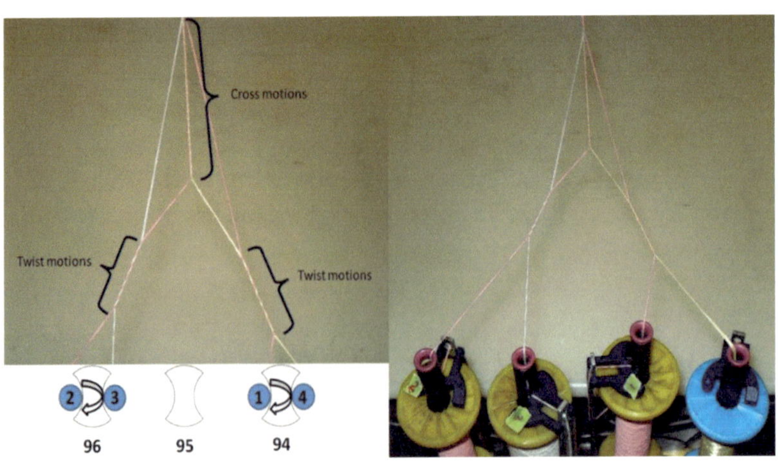

Figure 2. Basic stitches of lace.

http://dx.doi.org/10.5772/intechopen.82256

Although this unique nature of the Barmen lace industry had limited the dissemination of widespread knowledge, it did encourage the production of lace in the region. Barmen lace benefitted from the proximity of lace producers, lace machine designers and many technological developments are evident in the U.S. Patents issued to residents of Barmen and Wuppertal, Germany.

The Barmen lace machine is an evolution of the original mechanically geared braiding machine, often known as a Maypole braiding machine. In the Maypole braiding machine, the yarns are divided into two fixed groups of counter rotating directions producing two oppositely pitched sets of helices. The Barmen braiding machine allows individual yarns to change direction at effectively any point along its path. Similarly, as the Maypole braiding machine was inspired by the Maypole dance, the Barmen machine design inspiration comes from the agile hand motions of bobbin lace makers. The distinct advantage of the Barmen over the conventional Maypole braiding machine is found in the motion control of individual yarns. Pattern control in these machines is implemented using a Jacquard mechanism.

4.2. Barmen lace braiding machine

Figure 3 shows the general structure of a Barmen lace braiding machine. By comparison, this machine is significantly smaller than those other lace formation technologies. The basic components of the Maypole braider are also found in the Barmen braiding machine including frame, spur gear train, spindles, and take-up device. However, the Barmen machine employs more advanced features. The primary difference is the versatility of the driver plates (i.e., horn gears) which can be turned on and off as stipulated by operational rules. **Figure 4** is a schematic view of the top of a Barmen lace braiding machine. The even numbered driver plates turn clockwise and the odd numbered driver plates turn counter-clockwise. The even cycle must finish and the spindles or carriers stop before the odd cycle can begin.

Figure 3. Barmen lace braiding machine circa 1920 [14].

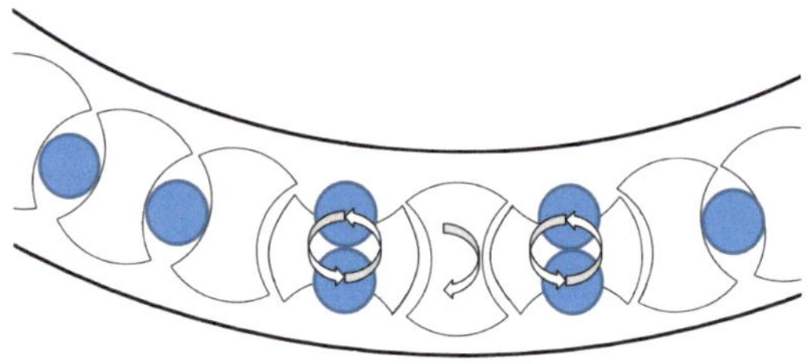

Figure 4. Schematic view of lace braiding machine.

Another notable difference of the Barmen lace braiding machine is the beat-up mechanism. This mechanism is akin to the weaving machine reed used to control fabric density. The beat-up mechanism is found in the center of the machine and consists of a dome with slits to allow reciprocating action of knife blades to compact the yarns following the corresponding beating motion. The blades are deployed as even and odd groups, according to the driver plate and spindle motions. These important advances over the Maypole braiding machine provide the ability to produce complex and irregular fabric structures.

The driver plates are positioned as a series of overlapping circles about the machine radius. The driver plate geometry is symmetric about two orthogonal axes with concave and convex regions. For a plate to rotate adjacent plates are required to remain stationary. Adjacent concave regions precisely permit the moving convex spindle cradle to pass without interference. This motion serves as the primary mechanism for imparting motion to the spindles and ultimately the yarn. In the same way as the traditional braiding machine, two different motions are required to pass a spindle.

4.3. Known materials used

Lace machines have been employed with a variety of materials, both natural and synthetic fibers, in the production of fancy lace and other apparel products. Marenzana [15] describes the use of Rayon fiber with lace braiding machines. Surface fiber treatments, known as sizing, may improve the lace braiding process as well as resin-fiber compatibility in subsequent composite manufacture. The use of high performance fibers such as those commonly found in composite materials has not been reported in the literature.

5. The modern lace braiding machine

The modern lace braiding machine is a direct descendent of the Barmen lace machines. The modern lace braiding machine has been continually improved; as witnessed in the numerous European, Japanese, and international patents. Some notable improvements include electromagnetic actuation of driver plates which allow electronic pattern control and computerized design to operate seamlessly without Jacquard punch paper. The electronic control eliminated the need for a mechanical Jacquard mechanism. Improvements in machine materials have increased the wear resistance and life of components. Certain mechanical features of the modern lace braiding machine protect the components and its lace product during production. For example, if the yarn breaks during production, the machine will automatically shut off as the bobbin carrier shorts an electrical switch so that the lace fabric can be saved and the broken yarn can be repaired by simply tying a knot. If this feature were not available, each time a yarn broke, the whole lace fabric would have to be discarded. A second feature is the construction of clutches out of a low-cost plastic material. If certain components fail to function perfectly, the clutch will fail before machine damage occurs. These inexpensive clutches can be replaced relatively easily and quickly. These two protective features are essential for industrial lace braiding but they limit the size and type of yarn that can be used.

Presently many of the modern machines are produced by Asian manufacturers who are in proximity to the textile manufacturing locations, although the Krenzler Company still manufactures lace braiding machines in Germany [16]. The machine evaluated in this research is manufactured in South Korea and clearly has its engineering origins from the Barmen lace machine. Considering improvements in the Barmen lace braiding machine during the last 30 years and the fact that many lace braiding machines are now manufactured outside of Germany, we refer to these machines as modern lace braiding machines. We acknowledge that the modern lace braiding machines originated from the Barmen lace machines.

Figure 5 is an engineering rendering of the modern lace braiding machine evaluated during this research. **Figure 6** illustrates the bobbin spindle actuation assembly. Spindle cradles are used to move yarns with the driving plates. The solenoid actuates the plastic cam which in turn lifts the plastic fork and plastic clutch and engages the driver plate allowing the spur gear to rotate the spindle cradles and perform the basic cross and twist motions on the yarns. After 180 degrees of rotation, the cam pushes against the inactive solenoid and a compression spring forces the fork and clutch to the resting position while spindle cradles and bobbins remain stationary.

5.1. Bobbin spindles (carriers)

Another important component is the bobbin spindle. Commonly referred to as carriers, they control the tension in the braiding yarn as well as allowing the release of new material during braiding. See **Figure 7** for the following operational details. The yarn is unwound from the bobbin and passed through an initial eyelet making a 90-degree bend where it continues until a second eyelet is located which also requires a 90-degree bend toward the spindle center where another 90 degree turn over a ratcheting pawl is required. The yarn now travels down the center of the spindle tube i.e. bobbin axis of revolution where a tension spring with eyelet requires a 180 degree turn. Finally, the yarn moves up the tube where a final ceramic eyelet allows the yarn to reach the fabric formation zone.

Figure 5. CAD drawing of modern lace braiding machine.

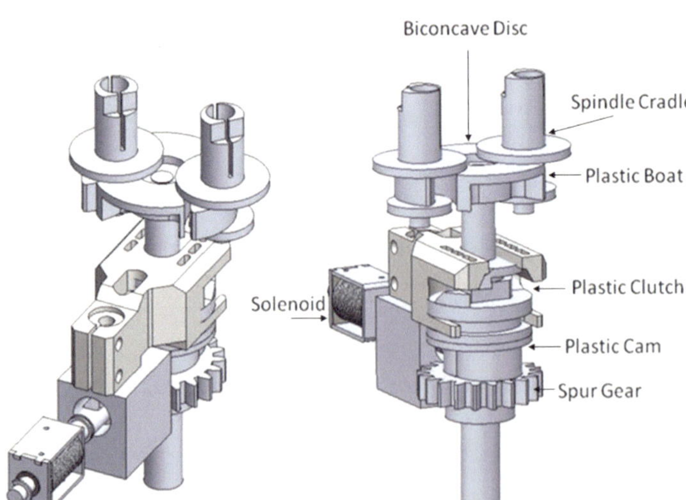

Figure 6. Main bobbin actuation assembly (front and rear views).

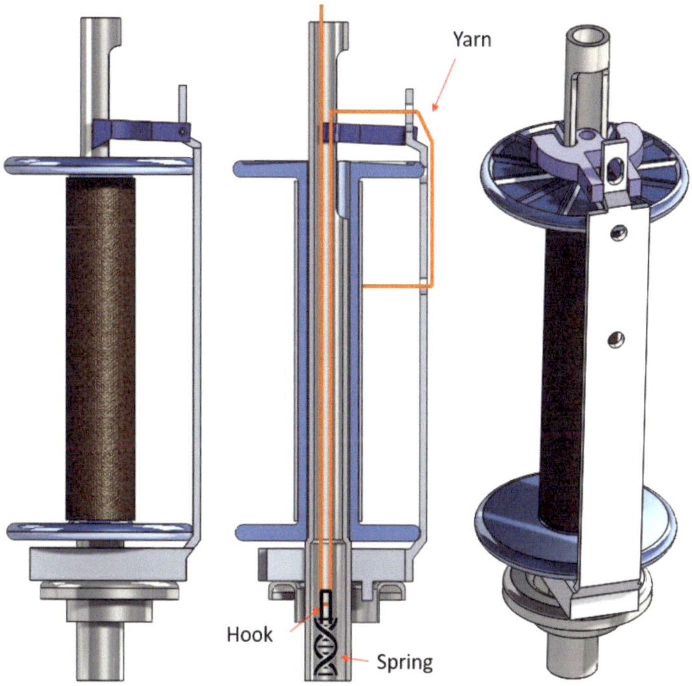

Figure 7. Bobbin spindle (carrier).

6. Analysis of machine for manufacture of structural composite pre-forms

Lace braiding technology has been demonstrated in the manufacture of intricate and decorative fabrics for more than a century. If lace braiding machines are suitable for handling large high strength yarns such as aramid and even carbon fiber prepregs, it was thought that the structures might be suitable for use as planar and 3-D space trusses. An evaluation of a modern lace braiding machine is performed on the typical execution to determine if braided composite strength-to-weight ratio could be improved by utilizing a lace braiding technology. A modern lace braiding machine incorporating a computer controllable electro-mechanical yarn interlacing system was purchased to test the proposition that it might be used to more efficiently orient and interlace yarns to create a truss-like pre-form in either a flat or cylindrical form [17]. **Figure 8** is an example of a CAD model for a proposed composite tube manufactured with a lace braiding machine.

Figure 9 shows the initial lace fabric preform made with a modern lace braiding machine during the evaluation and research phase of this work. The small white yarns are cotton. **Figure 10** shows a flat lace manufactured on a modern lace braiding machine made from larger twisted yarns.

Table 1 denotes a list of advantages of the modern lace and braiding machine.

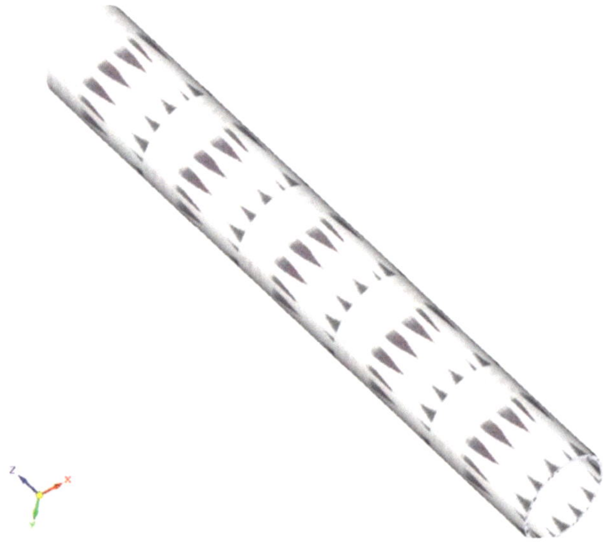

Figure 8. CAD model of lace braided composite tube.

Figure 9. Initial preforms evaluated for composite reinforcement.

Table 2 denotes a list of problems encountered with the modern lace braiding machine during the evaluation of this study and comparison to conventional Maypole braiding machines.

Figure 2 shown previously is an initial attempt to make a lace from high performance yarns (1100 denier). In this attempt, we discovered that the yarn carrier mechanisms supplied with the machine are not well suited for using larger yarns. Large and thus stiffer yarns are needed for producing lace pre-forms suitable for structural composite applications.

When large, high strength yarns (<2400 denier) such as Kevlar®, Vectran®, and carbon fiber were used, the clutches would quickly fail because tension developed in the yarns due the yarn

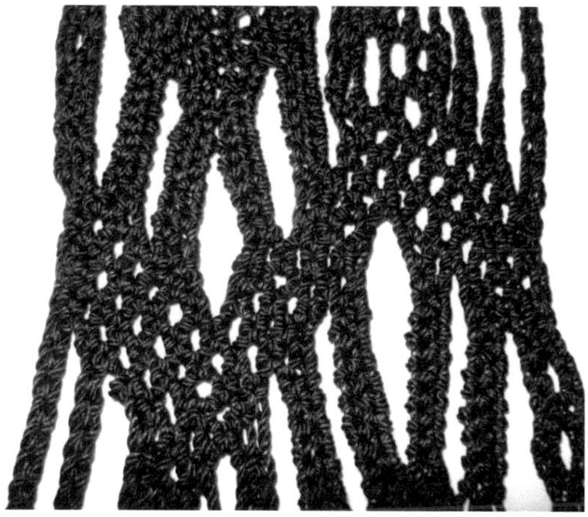

Figure 10. Machine braided lace.

General advantages of modern lace braiding machines
• Control of individual bobbins
• Small driver plates leads to increase in number of total yarns for given machine size
• Bobbin carriers are easily removed
• Cylindrical and flat fabrics produced from the same machine
• Precision fiber placement
• Open structure amenable to truss formation is possible
• Yarn crossovers are stabilized by a 360-degree twist around the adjacent yarn

Table 1. Advantages of lace braiding machine technology.

stiffness and breaking strength exceeded the capacity of the carriers and clutches. Furthermore, these yarns would not break if the machine had payout and tension problems. This would result in excessive clutch failure and machine down time. Constant clutch replacement is time consuming. After repeated adjustments to the machine, it was determined that the machine would not operate consistently with large, high strength yarn without the machine shutting down and/or breaking plastic clutches. Solving this problem will require a more robust machine design of the clutches and improved carrier payout necessary for braiding. In the process of evaluating the lace braiding machine, several other structural deficiencies were noted.

Despite these inherent limitations, the feasibility of using lace braiding technology was "proved in concept" when a more robust machine can be designed and built. To do this, some open structure lace patterns have been designed and produced using the light-weight yarns that the machine could process. Composite preforms using lace-like patterns possible with lace braiding have been made on a conventional Maypole braiding machine and evaluated for strength and stiffness to further promote the structural lattice concept [17].

Modern lace machine	Conventional Maypole braiding machine	Lace machine yarn carrier	Maypole yarn carrier	Lace fabric formation	Maypole fabric formation
Frequent clutch failure	No clutch-direct drive	Excessive yarn bending to payout	No clutch-direct drive	Fixed angle to fell point	Variable angle to fell point
Complex mechanism	Simplified mechanism	Poor tensioning for large HS yarns	Adequate tensioning of large HS yarns	Yarn fiber disintegration due to abrasion with the machine parts	Minimal abrasion with the machine parts
No axial yarns	Axial yarns	Small carrier/ bobbin capacity	Larger carrier/ bobbin capacity	Buildup of yarn twist	Minimal imparted yarn twist
Accumulation of debris in driver plates	Accumulation of debris in driver plates	Non-rotating terminal eyelet	Rotating terminal eyelets	Beat-up mechanism	No beat-up mechanism

Table 2. Problems encountered with lace braiding machine evaluated.

6.1. Other important issues: twisting of yarns, beat-up mechanism, and machine design

The carriers used in lace braiding are free to rotate about the plate. Motion about this additional degree of freedom will be exacerbated at high speeds as inertial effects increase and may potentially cause problems. The freedom of the carrier to rotate during braiding can cause excessive buildup of twist in the yarn. **Figure 11** illustrates an example of this phenomenon

Figure 11. Buildup of excessive twist in yarn.

Figure 12. Accumulation of broken fibers at fell point due to beatup knife abrasion.

due to carrier rotation. Twist build up in the yarns can be alleviated by pre-twist that counteracts the twist occurring in the opposite direction, however requires additional processing and reduces yarn stiffness. Careful consideration of the bobbin movements required by the yarn paths in each pattern can reveal the amount and direction of pre-twist required to eliminate the buildup. Alternatively, a rotating terminal eyelet can be employed to reduce yarn twist, else other means are required.

The braiding formation process can cause severe damage to yarns. The violent action of the beatup knife mechanism damages fibers. **Figure 12** illustrates an accumulation of broken fibers resulting from the formation process of lace braiding machines. Minimizing abrasion is especially important if high performance brittle fibers are to be utilized.

7. Conclusion

Lace braiding technology has been introduced and a brief historical context provided. A description of how the components function has been presented. The lace braiding machine components and their functionality were described to demonstrate how the machine works as well as to assess the limitations for producing structural scale composite preforms. Based on the experiments that were made on the modern lace braiding machine, the machine deficiencies for manufacturing composites (listed in **Table 2**) are discussed. Suggestions for remedies in the machine design and operation are presented to enable future progress to be built upon addressing the current limitations while further advancing the future of lace technology in new areas such as space and aerospace. Complete re-design and construction of a machine suitable for composite preforms were considered beyond the scope of the research and left for future work.

Acknowledgements

The authors would like to thank the Alabama Space Grant Consortium NASA Training Grant #NNG05GE80H for supporting this work. The authors would also like to thank Jeff Thompson, Kevin Horne, Austin Yuill, David Jackson, and the Auburn University Department of Polymer and Fiber Engineering Braiding and Composites Research Laboratory.

Conflict of interest

The authors have no conflict of interests to declare.

Author details

David Branscomb[1]*, Yang Shen[1], Vladimir Quinones[2], Royall Broughton[2] and David Beale[2]

*Address all correspondence to: davidbranscomb@yahoo.com

1 Highland Composites, Statesville, NC, USA

2 Department of Mechanical Engineering, Auburn University, Auburn, AL, USA

References

[1] TEF [Internet]. Available from: https://www.tefbraids.com/

[2] Branscomb D, Beale D, Broughton R. New directions in braiding. Journal of Engineered Fibers and Fabrics. 2013;**8**:11-24

[3] Branscomb D, Beale D, Brougthon R. Computer-aided product and process development of lace braided composites. ASME Early Career Technical Journal. 2010;**9**:215-220

[4] Kyosev Y. Braiding Technology for Textiles, Principles, Design and Processes. 1st ed. Oxford: Woodhead Publishing; 2014. DOI: 10.1016/C2013-0-16172-7

[5] Lee S. Handbook of Composite Reinforcements. 1st ed. New York: John Wiley and Sons; 1992

[6] Yang C. Research and melioration on the computerized lace braiding machine [thesis]. Wuhan: Huazhong University of Science and Technology; 2007

[7] Earnshaw P. Lace Machines and Machine Laces. 1st ed. Guildford: Gorse Publications; 1995

[8] Textile World Record. Boston: Lord & Nagle Company. 1909

[9] Blahedo. Bobbin Lace [Internet]. Available from: http://en.wikipedia.org/wiki/File:Ursuline_lace_2.jpg [Accessed: 2012-09-15]

[10] The British Museum. (n.d.). A History of the World [Internet]. Available from: http://www.bbc.co.uk/ahistoryoftheworld/objects/0JxzTILOQLKGBZehim601A [Accessed: 2012-09-17]

[11] Krenzler G. United States of America Patent No. 979,770 SN 534,061. 1910

[12] Krenzler E. United States of America Patent No. 946,445 SN 502,529. 1910

[13] Farrell J. Identifying handmade and machine lace. In: The Museum of Costume and Textiles, Nottingham on 21st February 2008 [Internet]. Available from: http://www.dressandtextilespecialists.org.uk/wp-content/uploads/2015/04/Lace-Booklet.pdf. [Accessed: 2012-09-15]

[14] Middleton G. Imitations of Handmade-Lace by Machinery [Internet]. Available from: https://www2.cs.arizona.edu/patterns/weaving/articles/nb38_ml1.pdf [Accessed: 2012-09-15]

[15] Marenzana R. Yarn limitations in manufacture of barmen Torchon laces—Use of rayon increasing. The Melliland. 1930;**2**:290-293

[16] Krenzler [Internet]. Available from: http://www.krenzler-gmbh.de/index.htm [Accessed: 2012-09]

[17] Branscomb D. Minimal weight composites utilizing advanced manufacturing techniques [thesis]. Auburn: Auburn University; 2012

Chapter 5

Polymeric Synthetic Fabrics to Improve Stability of Ground Structure in Civil Engineering Circumstance

Han-Yong Jeon

Additional information is available at the end of the chapter

http://dx.doi.org/10.5772/intechopen.81246

Abstract

Polymeric synthetic fabrics are continuous sheets of woven, nonwoven, knitted, or stitch-bonded fibers and yarns. The sheets are flexible and permeable and generally have the appearance of a fabric. Among polymeric synthetic fabrics, geosynthetics including geotextiles have special functions of separation, filtration, drainage, reinforcement, and erosion control in civil engineering applications. Also, geosynthetics such as geotextiles and geogrids are used in asphalt pavement reinforcement. An important function of these geotextiles is as cushion layers to prevent puncture of geomembranes (by reducing point contact stresses) from stones in the adjacent soil, waste, or drainage aggregate. Geotextiles, however, are made from a combination of two or more polymeric synthetic fabrics. In this chapter, geotextiles as polymeric synthetic fabrics are introduced not only for improvement but also maintaining stability of ground structure in civil engineering circumstance with their related technologies.

Keywords: polymeric synthetic fabrics, geosynthetics, geotextiles, special functions, cushion layers, stability of ground structure

1. Introduction

Geotextile is classified into woven fabric and nonwoven fabric in a morphological form and it performs functions such as reinforcement, separation, filtration, and drainage when applied to civil engineering structures. Generally, due to its structural form, the nonwoven geotextile has a small permeability coefficient and permittivity despite its small apparent opening size (AOS) compared with the woven geotextile style, it has advantages in function. Woven geotextile is applied to reinforce soil structure with poor shape stability in nonwoven geotextile based on excellent mechanical performance, and it also takes charge of filtration and drainage [1, 2].

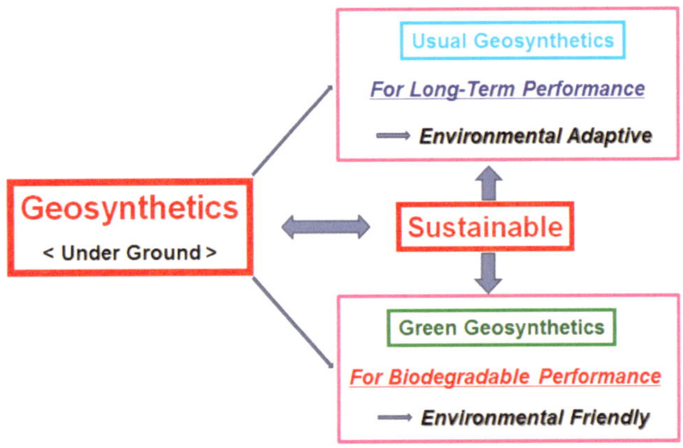

Figure 1. Sustainable geosynthetics.

Geotextile products have the characteristics of so-called tailor-made materials, which are known to function for specific applications. The long-term performance of geotextiles has a close relationship with the stability of the applied structure and practical applications such as continuous new method and new technology [3, 4]. As the demand and necessity to high-performance products gradually increase, composite products, environment-friendly products, environment adaptive products, hybrid, or smart products should be developed. In response to this, the development and advancement of the evaluation method are progressing steadily.

International Geosynthetics Society (IGS) Education Committee established that geotextiles can be classified broadly based on the manufacturing method and geotextiles is a continuous sheet of woven, nonwoven, knitted, or stitch-bonded fibers or yarns. Sheets are flexible and permeable and generally have a cloth-like appearance. Geotextile is used for separation, filtration, drainage, reinforcement, and anti-erosion applications [5].

In general, sustainable geosynthetics mentioned here are classified as "Usual Geosynthetics" and "Green Geosynthetics" based on required performance as shown in **Figure 1**.

In here, "Usual Geosynthetics" refers to the function-oriented long-term maintenance and environment-adaptive products introduced, and "Green Geosynthetics" refers to environment-friendly degradable geosynthetics, respectively.

In this chapter, we will introduce "Sustainable Geotextiles," which is differentiated from geotextile products to hybrid geocomposites except the traditional geotextile products.

2. Raw materials for geotextile products

The geotextile products and the polymeric raw materials are shown in **Table 1**. As the additives, internal fillers, antioxidants, carbon black, emulsions, and plasticizers are used for improving and complementing the physical properties, glass fiber, carbon fiber, aramid fiber,

Polymeric raw materials	Geosynthetic products
Polyethylene (Low, middle and high density)	Geotextiles, Geomembranes, Geogrids, Geopipes, Geonets, Geocomposites
Polypropyrene	Geotextiles, Geomembranes, Geogrids, Geocomposites, Prefabricated Board Drain (PBD)
Polyester (High tenacity)	Geotextiles, Geogrids, Prefabricated Board Drain (PBD)
Polyamide	Geotextiles, Geogrids, Geocomposites
Glass fibers, Polyvinyl alcohol(PVA) fibers, Aramid fibers, Carbon fibers etc.	

Table 1. Geosynthetic products with polymeric raw materials.

acrylic fiber, asbestos fiber, and low-modulus fibers such as polypropylene, polyamide, polyethylene, and polyester fiber, etc. are generally used to manufacture geosynthetic products.

Otherwise, antioxidants, carbon black, oils, plasticizers, fillers, etc. are added to improve the specific properties of the polymer, and two or more raw materials may be blended to improve specific properties. In the case of the geotextile made of polyethylene resin, radicals are formed due to sunlight, which causes decomposition and causes embrittlement. Hindered amine light stabilizers (HALS) series oxidation stabilizer is added to prevent radical formation by daylight and ultraviolet rays. Weather resistance is also improved. When geosynthetics are applied for a long period of time, durability depends on the characteristics of the polymeric materials used. Therefore, it is highly desirable to analyze the characteristics of geosynthetics to determine their use.

3. Fibers used for polymeric synthetic fabrics

3.1. Natural fibers

Natural fibers used in geotextile products are very limited, but they were first used as geotextile products. They were mainly applied in fiber, yarn, and knit form, and their demand increased as nonwoven- and woven-type products were developed. Since geotextiles of natural fibers have the advantage of being eco-friendly materials, the utility of geotextile products has recently begun to reappear. The raw materials of the products also include cotton, jute, coir, straw, and other stem forms of waste assembly, and it is very diverse. However, since it is not used much and cannot be mass-produced compared with synthetic materials, it poses a difficult problem to create demand. Some of them use civil engineering natural fiber products as slope stabilization, erosion control, drainage, etc.

3.2. Synthetic fibers

One of the conditions that geosynthetic products must have is economic advantages, which is a very real problem directly linked to manufacturing costs. Polyolefin, polyester, and so on are widely used as synthetic polymer fibers, and polyurethane, glass, and carbon-based polymers are applied to very limited fields in order to give a special purpose and function. Demand creation of geosynthetic products using polymer materials can be increased, and new functional products are expected to be developed in parallel with the development of various additives.

3.3. Recycled fibers

Since the fiber polymer materials used in the manufacture of geosynthetic products are often used in large quantities, therefore, the cost is low. Therefore, if the performance is similar, the manufacturing cost should be low. In view of this, in the case of nonwoven geotextile, products using already recycled polyester materials are being manufactured and sold, and interest and research on recycled polymeric materials are being actively pursued in terms of environment friendliness. However, in the case of the geosynthetic products manufactured using the recycled polymeric material, the physical properties are deteriorated, and therefore, there is a problem that it needs to be supplemented or improved in the future.

4. Manufacturing of polymeric synthetic fabrics as geotextiles

4.1. Geotextiles

Geotextile is a planar, permeable, polymeric (synthetic or natural) textile material, which may be nonwoven, knitted, or woven, used in contact with soil/rock and/or any other geotechnical material in civil engineering applications (**Figure 2**).

There are woven geotextiles which are divided into plain weave and twill weave using staple and filament yarns. Yarn used is usually as of 1000–3000 denier. And fabric density is generally in the range of 19–21 plies per inch in the warp and weft direction, and mainly polyester and polypropylene fibers are used, but polypropylene fiber has a weak light resistance. In addition, nonwoven geotextile, in which long fibers or short fibers are randomly arranged and bonded, is manufactured by using a needle punching and thermal bonding

Figure 2. Photographs of geotextiles.

process in the case of short fibers and laminated by spunbonding process in the case of long fibers in a weight of about 200–800 g/m^2.

In general, the constituent fibers form a disorderly entangled structure, so that they have excellent mechanical and mechanical properties, and polypropylene and polyester fibers are mainly used. Normally nonwovens are used for filter and separation functions. A nonwoven is a geotextile in the form of a manufactured sheet, web, or batt of directionally or randomly orientated fibers, filaments, or other elements, mechanically and/or thermally and/or chemically bonded. Nonwovens are used in filtration, drainage, separation, protection, and/or erosion control applications.

Fine soil particles can be captured in between the three-dimensional fiber entanglement of the nonwoven and prevent movement of these into the usually coarse "neighbor" soil. This way the buildup of a filter stable layer is possible. The geotextile filter can be dimensioned with available filter calculations [6, 7].

4.2. Geosynthetic clay liners (GCLs)

It is a geocomposite produced by bonding bentonite clay to a geotextile or geomembrane or filling bentonite clay between two geotextiles. The geotextile-made geotextile clay pottery often has a needle bent through the bentonite layer to increase internal shear resistance. It is effective as a barrier against liquids or gases when bentonite is hydrated. It is commonly used with geomembranes and is used as filler in landfills (**Figure 3**). GCL is also a factory-manufactured hydraulic barrier consisting of a layer of bentonite or other very-low-permeability materials supported by geotextiles and/or geomembranes, mechanically held together by needling, stitching, or chemical adhesives (**Figure 4**).

4.3. Geotubes and geocontainers

There are many opinions on how to prepare measures to be protected against or prevent catastrophic disasters such as tsunami and Katrina which have recently occurred, but one of the obvious ways of doing this is that it is closely related to advance prevention as well as disaster recovery. To do this, the method is the use of geotextile products. Geotextile containers, which are used instead of building rigid structures such as rocks and concrete in rivers, coasts, and harbors, are used as geotextile containers that are currently being used for this purpose worldwide, and they are used to construct flexible structures, and this technique has been successfully applied [8, 9].

Also, geotextile container is classified as geobags, geotubes, and geocontainers depending on the size and manufacturing method. The geotextile container is made by mechanically or hydraulically filling the soil including dredged soil in the geotextile bag. Generally, a geobag is a small geotextile container with a capacity of 0.3–5.0 m^3; it is usually used as a sand filling material, and it is finished with a small sewing machine.

Geotubes are manufactured in permeable geotextile and are filled with sand or dredged soil by hydraulic or mechanical methods. The diameter and length of the geotube depend on site conditions and installation possibilities, usually 150–180 m, 4–5 m wide, and 1.5–2 m high.

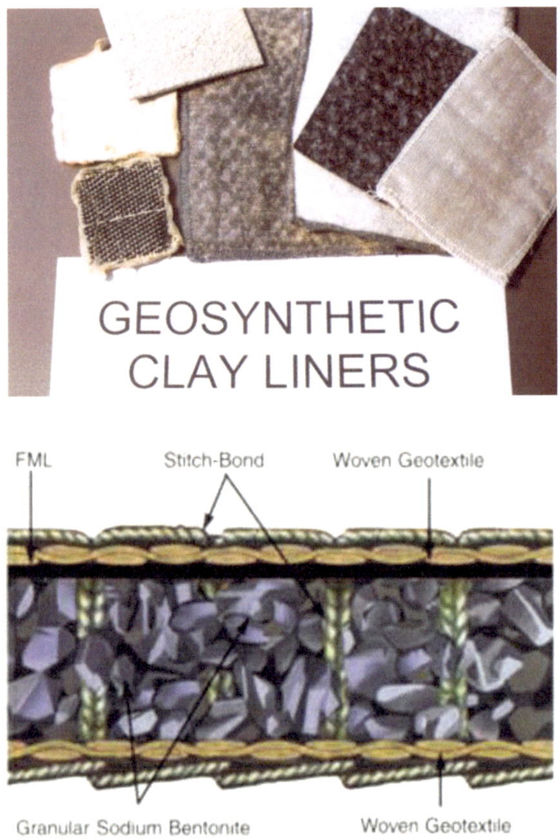

Figure 3. Photographs of geosynthetic clay liners.

In order to fill the upper part of the geotube with hydraulic method, the sandy soil should be closer (about 10 m) and the clayey soil as far as possible. Geotube is a massive pillow-shaped structure made in a permeable geotextile style and is filled mechanically with sand or dredged soil by a hopper or clamshell bucket (**Figure 5**).

Since the first attempt of geotube applications was in Brazil in the early 1980s, geotube application technology has been used as a containment embankment for the prevention and isolation of contaminated soil from France in 1986 and has since been used for underwater embankment or coastal protection in the Netherlands and Germany. Now, geotube was widely used for construction work [10].

Geocontainer is constructed by preliminarily sewing the geotextiles of the proper length together and installing it in the split bottom-dump width of the floor (the two ends are sewn together so as to form slender pillow shapes). And then, fill with sand or dredged soil, and seal the suture with a suture at the site (**Figure 6**).

The capacity of the geocontainer can be increased as the barge opening width of the barge becomes larger and is usually about 100–1000 m^3. When dredged clay is used, geocontainers

http://dx.doi.org/10.5772/intechopen.81246

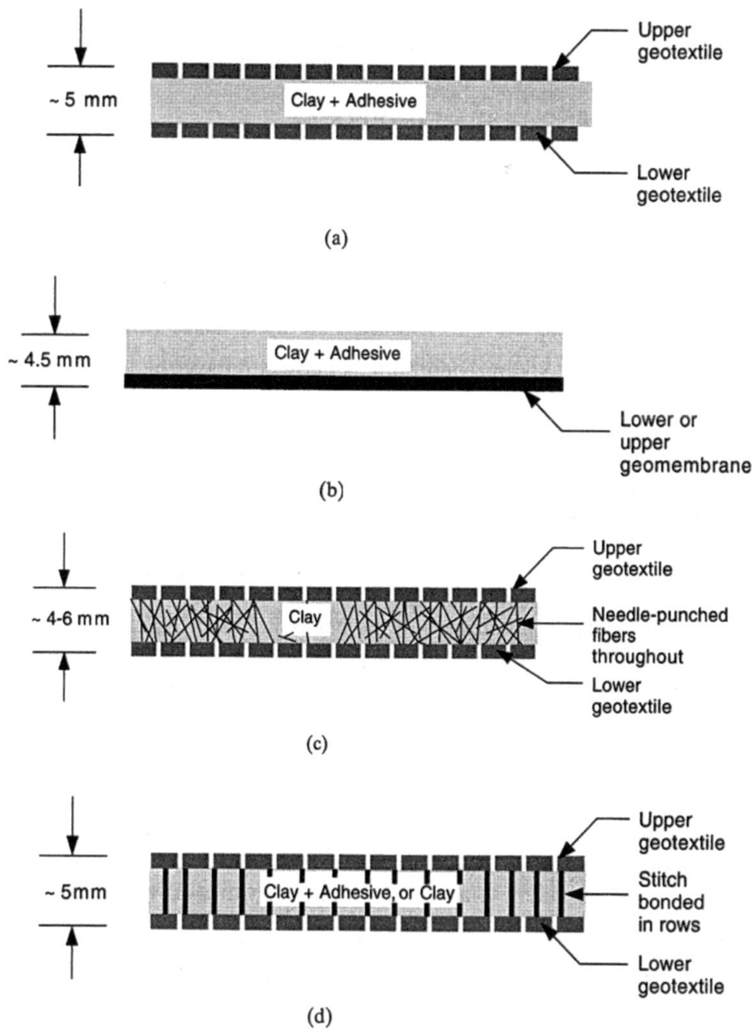

Figure 4. Cross-sectional sketches of currently available GCLs. (a) Adhesive-bound clay between upper and lower geotextiles, (b) Adhesive-bound clay above or below a geomembrane, (c) Needle-punched clay through upper and lower geotextiles, and (d) Stitch-bonded sketches of currently available GCLs.

can be manufactured by using nonwoven geotextile inside and woven geotextile outside. These geocontainers have many advantages such as shortening the installation period and reducing the construction cost due to the use of site-useable materials and workload and minimizing environmental pollution during construction.

Geocontainer application technology was first developed in the Netherlands and was used in 1986 in Germany for the construction of the flow-inducing dikes in the Rhine River and in 1987 in the Dutch-eroded canal's dikes.

The US Army Engineer Waterway Experiment Station (WES), which has recently been the centerpiece of the Army Engineer's Department and has been planned for Construction

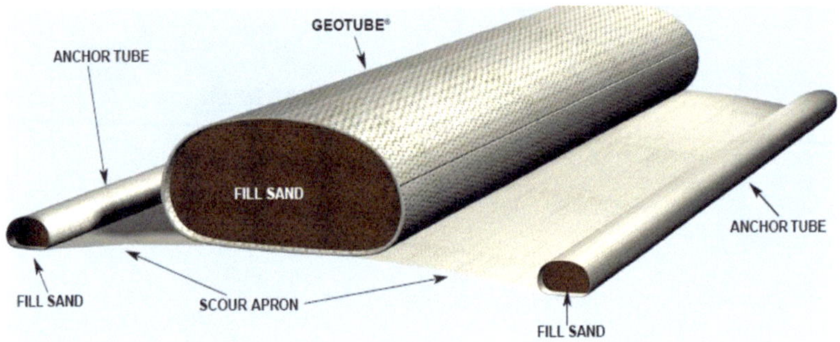

Figure 5. Schematic diagram of geotube.

Figure 6. Photographs of geocontainer application. (a) Spreading, (b) filling soils, and (c) dumping.

Production Advancement Research (CPAR) and has been developing innovative technologies using geotextile for the construction and maintenance of seawalls, rivers, canals, harbors, breakwater, dikes, coastal protection, roads, landfills, and reclaimed land.

5. Technical development trend of geosynthetics

Previously, environmental adaptive geosynthetics, which we have described as "Usual Geosynthetics," have not changed much over the past 20 years, but the paradigm of composite products using extreme strength fibers with the keyword of diversification of applications is being created. In other words, the development demand for divergence-targeted products, which means creation of usage as protection, maintenance, and restoration concept from natural disaster, is growing as megatrend of product development [11–14]. We will introduce the recently introduced fiber-reinforced geosynthetic products based on the concept in **Figure 1**.

5.1. Environmental adaptive geosynthetics

Environmental adaptive geosynthetics introduced as "Usual Geosynthetics" has not changed much over the past 20 years, but the paradigm of composite products using extreme strength fibers with diversification of uses has been created.

On the other hand, most of the synthetic polymeric materials that have been widely used are polyolefin-based and polyester-based ones. However, polyurethane, glass, and carbon-based polymers could be used to manufacture for special purpose and functions. Since the polymer materials used in the manufacture of geosynthetic products are often used in large quantities, therefore, the cost is low. Therefore, if the performance is similar, the manufacturing cost must be low.

In view of this, products using recycled polyester materials have already been manufactured and sold, and interest and research on recycled polymer materials are being actively pursued from the viewpoint of environmental friendliness. However, in the case of the geosynthetic products manufactured using recycled polymeric materials, the physical properties of the recycled polymeric materials are deteriorated, so that they have to be supplemented or improved in the future.

Recently, as the demand of composite-type geosynthetics has increased, functional and special high-performance materials have been used to improve the field application of geosynthetic products and to improve the stability of geotechnical structures from earthquakes, tsunamis, etc., liquid crystal polymer (LCP), polybutylene oxide (PBO), polypropylene sulfide (PPS), and meta- and para-aramid fibers have been used to combine with fusion technology for the production of hybrid geosynthetics.

5.2. Environment-friendly geosynthetics

"Green Geosynthetics" refers to products that have sustainable degradable geosynthetics and environmental pollution prevention and restoration functions that do not mean long-term implementation of initial performance in terms of environmental friendliness. In the case of geotextiles, "biodegradability" refers to a phenomenon in which initial performance is gradually lost over time due to decomposition by microorganisms or bacteria in the soil, which is a geotechnical structure. In terms of restoring the polluted environment, it is also a new area of geotextiles that meets the issue.

In order to manufacture "Green Geosynthetics," a resin which is biodegradable as a raw material should be used separately, and it is closely related to the reduction factor required for long-term use. Therefore, if the green geosynthetics is used as a filter, the production of a geotextile in the form of nanofibers will help improve filtration efficiency.

6. Development trend with geotextile-related products

6.1. Geotextiles

1. Nonwoven geotextile products

    High weight, over 5000 g/m^2

    Smart fusion multifunction product

    Filter products for nanofiber applications

    Composite products, etc.

2. Woven geotextile products

    High strength, 30 ton/m or more tensile strength demanded

    Creep property improvement product

    Low-elongation high-strength yarn use

    Smart fusion multifunction product

    Composite products, etc.

6.2. Geosynthetic clay liners (GCLs)

Differentiated hydraulic function product

Salt water swelling improvement product

Products with improved freeze-thaw stability

Selective-order function products, etc.

6.3. Filter and drainage geotextiles

Minimization of penetration by constraint load

Clogging prevention and minimization products

Biodegradable multifunctional products, etc.

6.4. Geotubes and geocontainers

High strength, 50 ton/m or more tensile strength demanded

Creep performance improvement products

Permeability and sealing property improvement products

Ultraviolet and salt water stability improvement products, etc.

6.5. Miscellaneous

Concrete reinforcement geocomposites

Silt fence products

Seam properties improvement products

Ultraviolet and salt water stability improvement products, etc.

7. Functional geotextile-related products

7.1. For separation, filtration, and reinforcement functions

In order to improve the separating function of the geotextile for reinforcement, it is possible to improve physical properties and permeability by designing the smoothness of the woven fabric at a high level and to improve the morphological stability by designing the tissue for controlling apparent opening size (AOS) [11–13]. Especially, it is designed to improve the tensile strength of fabric by improving density of weft yarn and double yarn design so as to improve the tensile strength in weft direction (**Figure 7**).

This product has the overall performance (chemical stability, higher tensile property, and water permittivity, etc.) as the geomembrane protection mat in the landfill construction caused by the working vehicle and the aggregate applied to the leachate drainage layer and at the upper part and can be used as a composite product.

7.2. Multiaxial geocomposite for reinforcement

As shown in **Figure 6**, geocomposite fabrication technology and products were developed to enhance the reinforcement function of geosynthetics by applying multiaxial knit fabric and geotextile composite technology by developing not biaxial but multiaxial knit. In addition, a smart monitoring high-performance multiaxial geocomposite technology is being developed in parallel to embed an optical fiber sensor in a multiaxial geocomposite appropriately to monitor the damage of the geocomposite due to stress concentration in real time (**Figure 8**).

7.3. Geotextiles for preventing reflective crack

Geotextiles applied on the top of the packed and unpacked road subgrade is considered to be the top layer of the bottom layer consisting of roadbed soil and the top layer consisting of soil or aggregate laid for construction. If the two layers are not properly separated, the particles of the lower layer penetrate the upper part, or the particles of the upper part penetrate the lower layer, causing settlement or cracking of the road. Also, when the bedrock is saturated by rain or other conditions, excess pore water pressure is generated by the traffic volume, so that the bedrock is weak and easily broken. Therefore, proper water discharge must be achieved, and

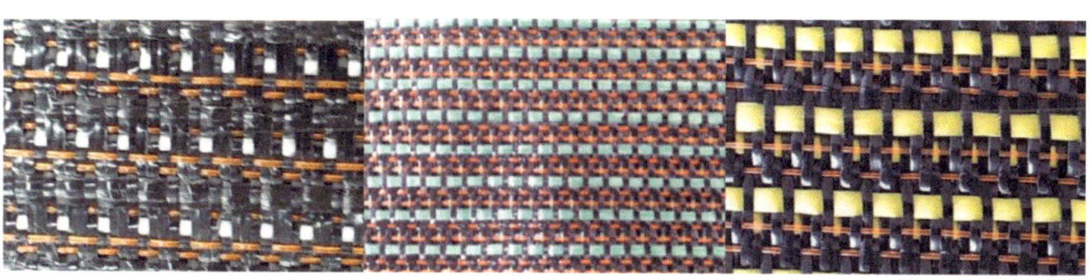

(a) (b) (c)

Figure 7. Geotextiles for separation, filtration, and reinforcement. (a) Separation, (b) filtration, and (c) reinforcement.

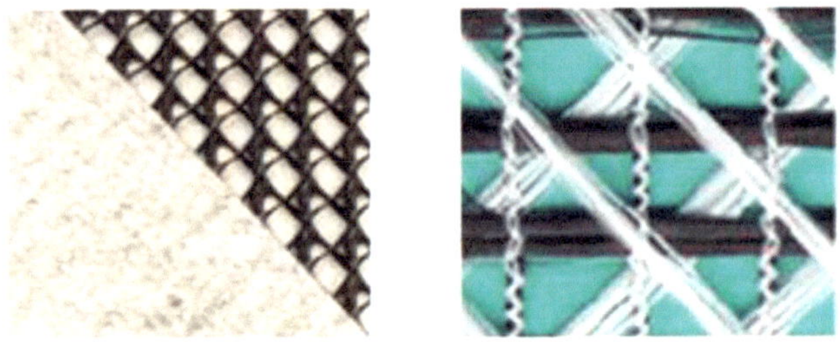

Figure 8. Multiaxial geocomposites for reinforcement.

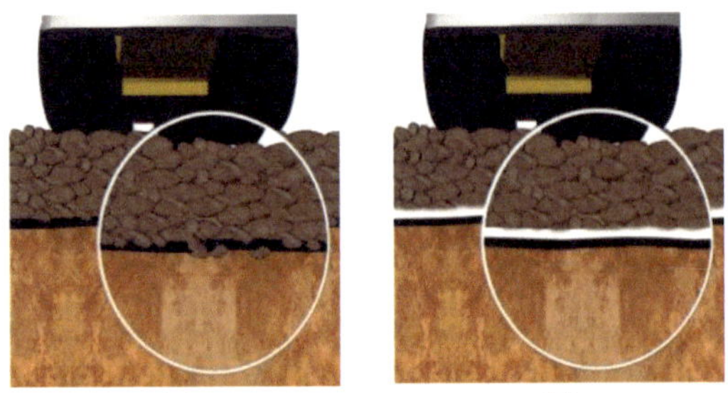

Figure 9. Failure with/without geotextile for pavement protection. (a) Without geotextile and (b) With geotextile.

proper pore size and good permeability coefficient are required because the piping phenomenon is required to prevent the loss of soil along with the flow of water [14, 15].

As shown in **Figure 9**, pavement roads are damaged due to cracks and plastic deformation before the design life due to the surrounding environment and repeated traffic loads, which causes wasted budget for maintenance of road pavement.

This is due to the weakening of the bearing capacity of the pavement ground or the cracking and growth due to the expansion and contraction of the water inside the packed asphalt or concrete. The role of a geotextile as a localized stress reduction layer could be to prevent or reduce damage to a given surface or layer by vehicle passing load.

Therefore, in order to improve the durability of the pavement, development is underway to improve the performance of asphalt or aggregate as a road pavement material and to reinforce the pavement by adding reinforcement materials such as geosynthetics to traditional pavement materials.

On the other hand, asphalt pavement using geotextiles has a great effect on prevention of fatigue cracks and reflective cracks and additionally has an advantage of blocking water penetration due to road crack by increasing water penetration. Advanced geotextiles have been developed, have improved toughness against repeated fatigue loads, and are resistant to various damage loads that occur during the construction process (**Figure 10**).

7.4. Biodegradable geotextiles

In the case of geosynthetics for slope reinforcement or erosion prevention considering vegetation, biodegradable products are required for the purpose of activating the planting of plants.

http://dx.doi.org/10.5772/intechopen.81246

Figure 10. Various geotextiles for antireflective crack propagation.

However, as mentioned above, even though it is a product of very important issue in terms of being environmentally friendly, it is easy to enter the market only if the stability of raw material supply and supply and product standardization are solved. Here, only the biodegradable vegetation mats and geocells used for vegetation in river maintenance and slope greening are introduced.

As shown in **Figure 11**, the geotextiles for vegetation mats have a very high initial dependency for the purpose of preventing or stabilizing the erosion. Therefore, biodegradation occurs in the course of the planting process after the vegetation mat construction, thus contributing to the improvement of the stability of the structure.

The synthetic resin system used for slope protection and erosion prevention was originally a product using a mat made of a heat-sealable webbing structure using nylon and a product with a reinforcing material (geogrid) combined with a web structure. And polypropylene staple fibers have been developed in the future, but since they are nondegradable products, it has been pointed out that the residues become an environmental pollution source after completion of the desired slope protection and erosion prevention function.

It is now in the process of restoration of various floods due to increasing weather conditions, eco-friendly construction methods, and landscaping and greening. As the demand for

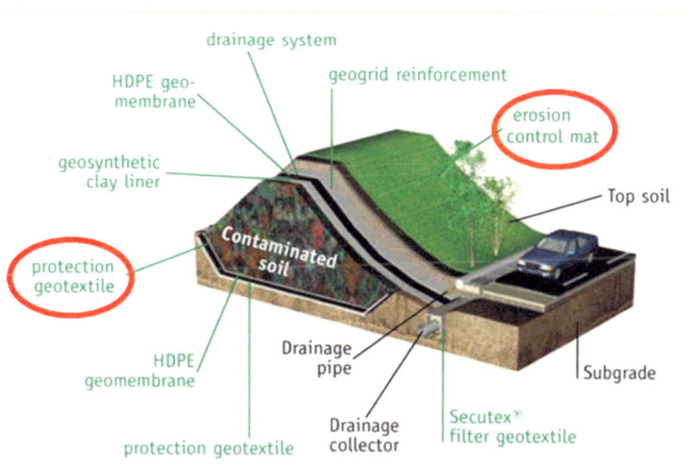

Figure 11. Application examples of geotextiles for erosion control.

products becomes greater, it is possible to apply and expand key technologies for vegetable mats made from biodegradable resins.

8. Nanofiber-used geotextiles

In general, geotextiles can be fabricated with a fiber size of more than 1 denier. However, when the size of a fiber becomes micro fine or nanofiber size, it has a great advantage in restoring the environment from pollution or improving filtration performance (**Figure 12**).

As shown in **Figure 13**, the filtering capacity of the geotextile depends on the number of fibers per unit area, the size of the pores, and the compositional structure. Therefore, when nanofibers are used, the smaller the pores constituting the geotextile, the removal rate of the toxic water is improved. However, at present, there is not a variety of techniques for manufacturing nanofibers, and since the manufactured nanofibers are expensive, the practical use of nanofibers is very slow.

In general, regular fibers are widely used to manufacture geotextiles and geogrids, but filtration efficiency of microfiber and nanofiber geotextiles is better than regular fiber-used geotextiles. To consider this, it is expected that nanofiber geosynthetics could be the smart filtration function in geoenvironmental applications by their composition structure as in **Figure 3**. If the numbers of filled fibers per unit area are increasing, pore size among nanofibers is decreasing. Therefore, the fine particles cannot pass through pores by nanofibers, and the filtration efficiency will be improved. This means that ultrathin geosynthetic filter can be manufactured with high-quality filtration function to absorb the fine impurities and toxic components in water and air media (**Figure 14**).

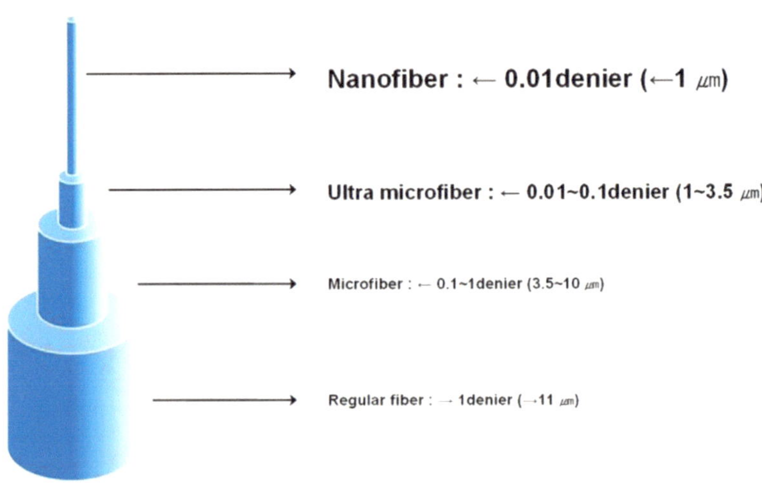

Figure 12. Thickness of fiber for geosynthetic fiber production.

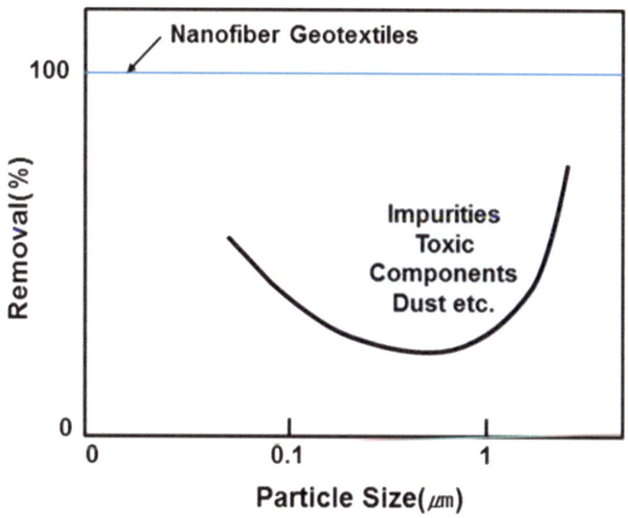

Figure 13. Impurity removal ratio according to geotextile pore size.

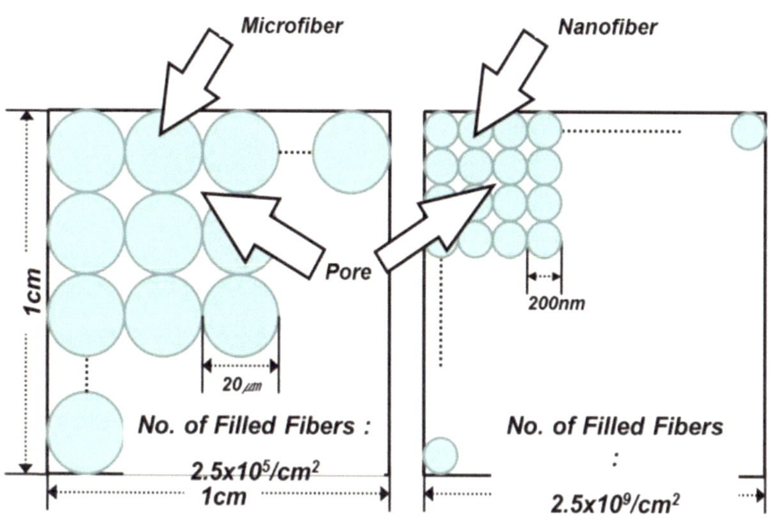

Figure 14. Geotextile filtration ability according to the number of filled fibers.

Figure 15 shows the separation concept of nanofiber air filter by pressure. To be the best air filter, higher particle collection and dust retention rate should be required.

In order to remove the heavy metals and toxic substances contained in polluted soil, nonwoven geotextile is used which is made by mixing nanoparticle clay with polyester fiber (**Figure 16**). Of course, the engineering performance of mixed nonwoven geotextile will vary depending on the composition of clay and particle size, but the strength degradation due to leachate, chemical, and biological degradation of waste landfill is not greater than that of

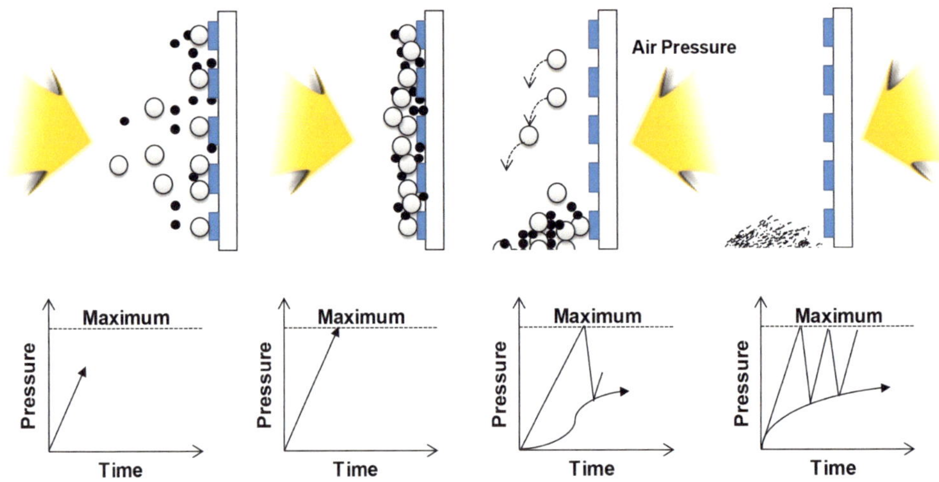

Figure 15. Maintenance of filtration efficiency for nanofiber filter.

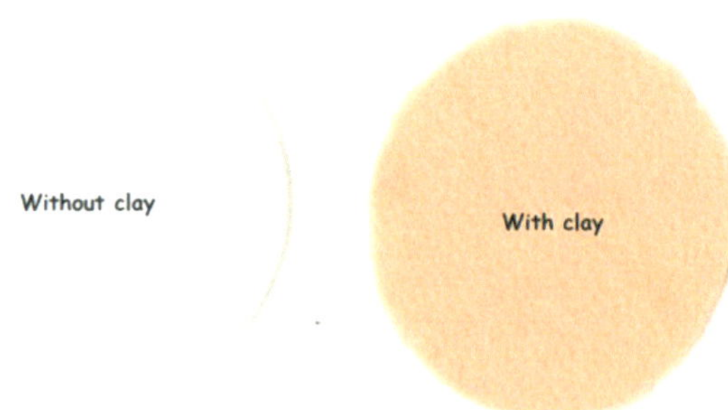

Figure 16. Nonwoven geotextiles with/without nano-clay.

nonwoven geotextile without clay. Also, AOS is higher than that of the nonwoven geotextile which is not mixed with clay, so that the permeability is improved.

9. Conclusion

Looking back at the civil engineering industry, product development and construction technology have been growing remarkably. There is also a growing demand for sustainable civil engineering products to protect, repair, and restore structures after recent floods, tsunami, and earthquakes. Considering the product characteristics and functions according to the product material and manufacturing method, civil engineering is rapidly growing with advantages of development of convergence composite geosynthetics using polymeric material, new design with geosynthetics.

In addition, it is expected that the utilization of civil engineering products will be further enhanced by various applications of the development and manufacturing methods of geosynthetics. For this, new convergence type composite geotextile-manufacturing technology should be developed not only standardization and reliability of evaluation methods but also design and construction methods and equipment.

Author details

Han-Yong Jeon

Address all correspondence to: hyjeon@inha.ac.kr

Department of Chemical Engineering, Inha University, Incheon, South Korea

References

[1] Giroud JP. Geotextiles and Geomembranes—Definitions, Properties and Design. Vol. 1-3. St. Paul, Minnesota: IFAI; 1984

[2] Koerner RM. Designing with Geosynthetics. 5th ed. New Jersey, USA: Prentice-Hall Inc.; 2005

[3] Rawal A, Shah T. Geotextiles: Production, properties and performance. Textile Progress. 2010;**42**:181-226

[4] Koerner RM, Koerner GR. Lessons learned from geotextile filter failures under challenging field conditions. Geotextiles and Geomembranes. 2015;**43**:272-281

[5] International Geosynthetics Socoity (IGS), Education Resource. https://www.geosyntheticssociety.org/education-resources/

[6] Koerner RM, Koerner GR. Geotextile filter failures: Examples and lessons learned. In: Proceedings of the 25th Central Pennsylvania Geotechnical Conference; Hershey, PA, (on CD). 2011

[7] Maiser M, Myles B. Possible culpability of filter geotextile in the failure of a sea wall. In: Proceedings of the 1st Pan American Geosynthetics Conference; Cancun, Mexico. 2008. pp. 833-841

[8] Lawson C. Geotextile containment—International perspectives. In: Proceedings GRI-17. Hot Topics in Geosynthetics IV, GII; Folsom, PA. 2003. pp. 178-201

[9] Heibaum M. Special issue on geotextile containers. Journal of Geotextiles and Geomembranes. 2002;**20**(5):279-342

[10] Hsieh CW, Lin CK, Chiu YF. The strength properties of geotextiles in ocean environments. In: Proceedings of the EuroGeo3; Munich, Germany. 2004. pp. 377-382

[11] Guglielmetti JL, Koerner GR, Battino FS. Geotextile reinforcement of soft landfill process sludge to facilitate final closure. Journal of Geotextiles and Geomembranes. 1996;**14**(7-8):377-392

[12] Alexiew D, Brokemper D, Lothspeich S. Geotextiles encased columns (GEC): Load capacity, geotextile selection and pre-design graphs. In: Proceedings of the GeoFrontiers 2005; Austin, Texas, (on CD). 2005

[13] Cuelho E, Steve P, Zachary M. Relative Operational Performance of Geosynthetics Used As Subgrade Stabilization, No. FHWA/MT-14-002/7712-251. FHWA, U.S. Department of Transportation, USA. 2014

[14] Suits LD, Koerner GR. Site evaluation/performance of separation geotextiles. In: Proceedings of the Geosynthetics '01 Conference. IFAI Publ.; 2001. pp. 451-468

[15] Perkins SW, et al. Geosynthetics in pavement reinforcement applications. In: Proceedings of the 9th ICG; Guaruja, Brazil. 2010. pp. 115-164

Chapter 6

Behavior of Reinforced Soil Wall Built with Fabrics

Mario Riccio and Mauricio Ehrlich

Additional information is available at the end of the chapter

http://dx.doi.org/10.5772/intechopen.79239

Abstract

This chapter presents an example of use of fabrics in geotechnical engineering construction. Some aspects related to design, construction, and the performance of a 4.2-m-high-reinforced soil wall, located in Brazil, is presented. In this wall, geogrid (fabric reinforcement) was used as reinforcement, and the backfill was a fine-grained residual tropical soil. The wall was monitored during its construction (2 months); load in the reinforcements, vertical and horizontal displacements of the reinforced soil mass, and efforts on block-face were measured. The monitoring of the wall was done by means of load cells for the reinforcements and block-face, and also includes settlement plates, total pressure cells, inclinometers, and topographical marks. The results provided by the instruments showed good performance of the wall. Measurements and calculated tension in the reinforcements were compared, and good prediction capability of the used analytical method was demonstrated. The measured tensile load in the reinforcements was lower than the admissible load of the geogrids used in the wall. Measurements also indicate that the block-face was able to support part of the load that would be carried by the reinforcements.

Keywords: fabrics, reinforced soil wall, monitoring, analytical method

1. Introduction

Reinforced soil walls (RSW) are retaining structures composed by facing, compacted backfill and usually geosynthetic reinforcements. Compacted soils have good strength in terms of compression solicitation, but they have a very low tensile strength. Thus, similar to the reinforced concrete, the use of fabrics as reinforcement is intended to provide enough tension resistance to the composite material. RSW structures can be built with a wide variety of fabrics (geosynthetics). Those fabrics are specially developed and have different applications in geotechnical engineering. **Figure 1** shows some examples of geosynthetics used in RSW construction as reinforcement.

(a) (b) (c)

Figure 1. Examples of geosynthetics used for RSW construction as reinforcement: (a) geogrid; (b) nonwoven geotextile; (c) woven geotextile.

The backfill used for reinforced soil wall construction could be purely sands or even soils that contain high percentage of fines. In Brazil, due to the abundance of residual fine-grained soils, it is a common practice to build RSW using this kind of soil. This kind of soils, in spite of its high percentage of fines, has high strength resistance, presents good workability, and achieves a proper density during compaction. **Figure 2** shows the basic concept of RSW; the geosynthetics link the active zone (the unstable zone) to the resistant zone. Design should provide enough reinforcements in order to guarantee no failure or pullout of reinforcements from the resistant zone. Both zones liked together works like a block that may be considered as a conventional retaining wall that provides the stabilization of the nearby nonreinforced soil mass. The mobilized load along the reinforcements is variable, and the location of the points of maximum tension defines the potential failure surface that separates the active and passive zones. **Figure 2** also indicates the shape of the potential failure surface that varies with the stiffness of reinforcements.

The design of an RSW comprises basically two verifications: (a) external stability that is basically the same concept used for the conventional retaining walls, i.e., stability analyses for sliding and

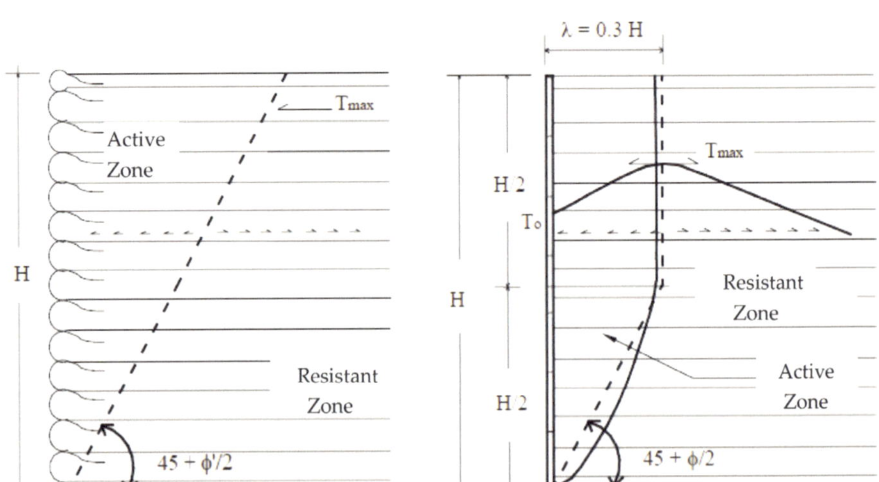

(a) (b)

Figure 2. Basic concept of RSW and the potential failure surface: for extensible (a) and rigid (b) reinforcements.

overturning, bearing capacity and general failure and (b) internal stability. The internal stability consists in the comparison of the mobilized load in the reinforcements (geosynthetic) with the tension strength of those ones. There are some methods to evaluate the mobilized load in the reinforcements, such as [1–5]. Through case studies, field instrumentation, physical and numerical modeling [6–11] have been demonstrated that among these methods the more suitable are the ones proposed in [4, 5]. These methods explicitly consider soil and reinforcement properties, the effect of compaction operation, and the relative stiffness between soil and reinforcement. The method described in [5] is based on the one developed by Ehrlich and Mitchell [4]; this method uses simple equations and may take in the calculation facing inclination into consideration.

Figure 3 shows different concepts of facing elements. In the RSW structures, facing has a secondary function, and it is used to avoid erosion and localized soil failure near to the face, besides providing suitable visual appearance. Precast concrete block-face is usually used in RSWs with geogrid reinforcements (**Figure 3c**). Precast concrete block-face is also used in the case of RSW with geosynthetic wrap-around facing (**Figure 3a**). This block-facing is applied after the end of the wall construction, and it is needed to protect geosynthetics from degradation due to exposure to ultraviolet rays and vandalism. Depending on its rigidity, the face may be capable to absorb part of the tension that would be supported by the reinforcements. Nevertheless, the design of internal stability is usually done without consideration of the facing contribution to the global stability, if it exists. Note that this approach is by the side of safety [6]. Moreover, enough drainage must be employed in order to guarantee no positive

Figure 3. Typical facing elements: (a) geosynthetic wrap-around facing before protection application (courtesy: Ober geosynthetics); (b) precast-concrete panels (courtesy: Reinforced Earth Company); (c) precast-concrete blocks facing; and (d) steel mesh facing filled with stones (courtesy: Paulo Brugger).

pore-pressures inside the reinforced soil mass. The drainage system is often composed by a vertical layer of gravel behind the face and a horizontal layer at the RSW bottom.

2. The São Jose dos Campos RSW

This section describes and shows monitoring results of an RSW built in the year of 2006, as a part of a road construction in the city of Sao Jose dos Campos, state of Sao Paulo, Brazil [6]. This RSW has 4.2 m height, segmental concrete blocks composing the face, and geogrid as reinforcements and tropical fine-grained lateritic soil as backfill. In the field, the soil compaction was done through a heavy vibratory roller drum Dynapac CA250PD. Other previous studies have also ensured good mechanical behavior of RSWs where fine-grained soil was used as backfill [12–18]. The wall under consideration was extensively instrumented during 2 months (constructive period) to verify its overall performance. The instrumentation consisted of load cells for measurement of the mobilized loads in the reinforcements and block-face, settlement plates, total pressure cells, inclinometers, and topographical marks. The main results obtained are presented and discussed in this chapter. The instrumentation indicates good mechanical performance of the RSW. The wall under analysis has not indicated any structural problems or excessive deformations. In Section 3, some design considerations and comparison of measured load in the reinforcements and predictions are shown.

2.1. Overall characteristics of the Sao Jose dos Campos RWS

In the wall construction, two residual soils were used as backfill, both with high percentage of fines. The yellow sandy clay (soil A) was used from the top of the wall to the 3.2 m depth, and red sandy clay, from 3.2 m depth to the bottom of the wall. In **Table 1**, the grain-size distribution and Atterberg limits (liquid limit, w_L, and plasticity index, PI) of those soils are presented. Using the Unified Soil Classification System, both soils were classified as CL (low-plastic clays).

Those backfill soils were tested in laboratory by means of plane strain tests. The plane strain condition is representative of typical wall behavior where the longitudinal length of the wall is much greater than its height. Under these conditions, it is a reasonable assumption the consideration of the absence of longitudinal deformations. The soil specimens used on tests were compacted statically with the same unit weight (γ) and water content (w) verified in the field. In **Table 2**, the results of those tests are shown; where ϕ is the friction angle of the soil (total stress envelope); c is the cohesion of the soil (total stress envelope); n, k (for loading),

Soil	≤2 μm (%)	≤20 μm (%)	≤2 mm (%)	w_L (%)	PI (%)
A	42	49	99	38	22
B	42	47	99	49	29

Table 1. Soil grain size distribution and Atterberg limits.

Soil	γ (kN/m³)	w (%)	ϕ (°)	c (kPa)	n	k	k_u	Rf
A	16.7	20	36	60	0.47	392	588	0.86
B	16.7	20	38	50	0.36	566	849	0.95

Table 2. Results of plane strain tests performed on the backfill soils.

k_u (for unloading), and R_f are hyperbolic parameters obtained from the triaxial tests according to the procedure followed in [19]. In the absence of plane strain or triaxial tests, the values of n and k can be selected using the suggestion from [20]. The value of k_u can be considered as 1.5 k, and R_f equals to 0.90 as typical values.

Two different PET geogrids were used in this RSW as reinforcements. One was placed in the reinforcement layers 1–3 (bottom to top) and the other in the layers 4–7. In **Table 3**, the characteristics of those fabrics are shown. In **Table 4**, the characteristics of blocks used as facing are also presented. The blocks were filled with crushed stones, in order to increase the pullout resistance of the geogrid-blocks interface and guarantee drainage at the face.

2.2. Instrumentation

Figure 4 shows a general view of the wall just after the end of construction. In **Figures 5** and **6**, are shown a cross section and plan view of the wall with the location of the instruments used for monitoring, respectively. The wall has seven layers of reinforcements with 3 m length each. Four of those layers were instrumented, i.e., reinforcement layers 1, 4, 5, and 6 (see **Figure 5**).

Reinforcement layers	1–3	4–7
Ultimate longitudinal tensile strength (kN/m)	55	35
Ultimate transverse tensile strength (kN/m)	30	20
Elongation at rupture (%)	12.5	12.5
Weight (gf/m²)	360	210
Opening size (mm)	20 × 30	20 × 20
Stiffness modulus, J (kN/m) at 5% strain	400	260

Table 3. Physical and mechanical properties of the fabrics (geosynthetic).

Dimensions (m)	0.2 height, 0.40 long, 0.40 wide
Block weight (kgf)	29
Block with*crushed stone (kgf)	40–50
Compressive strength (MPa)	6–12

Table 4. Characteristics of concrete block used as facing.

Figure 4. General view of the RSW just after construction.

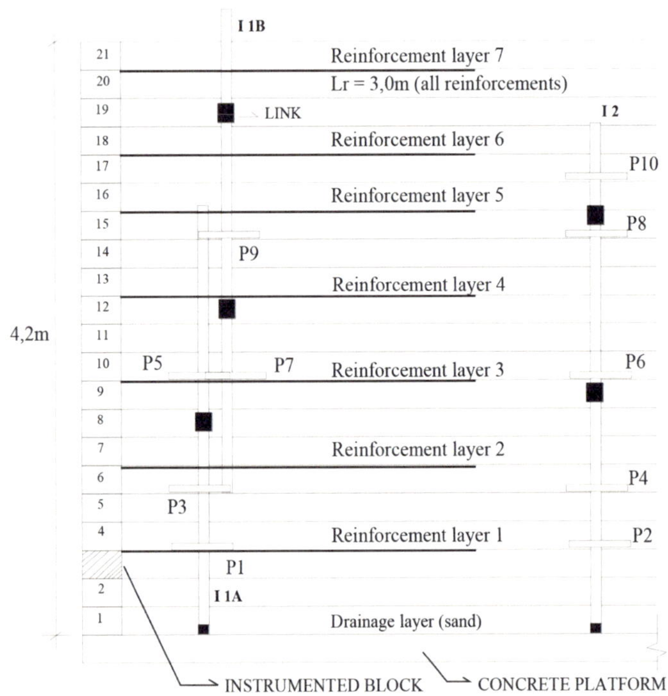

Figure 5. Cross section of instrumented wall: P is settlement plate and I is inclinometer, [6].

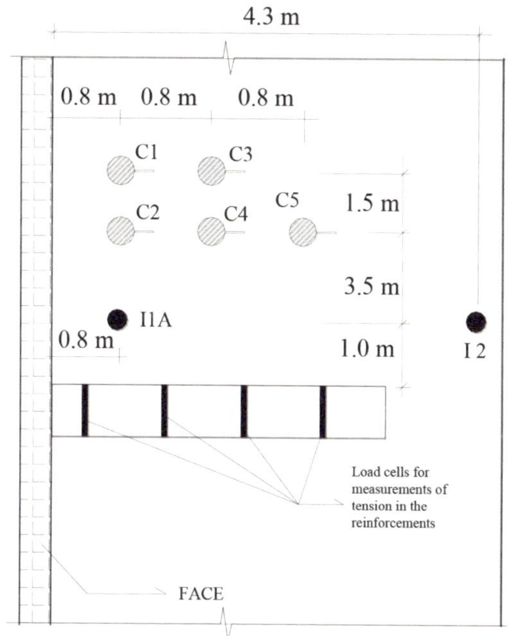

Figure 6. Location of the instruments in the first layer of reinforcement at 3.6 m depth, [6].

Inclinometers (I1A, I1B, and I2) and magnetic settlement plates (P1–P10) were used to measure lateral and vertical movements, respectively.

Topographical measurements were used for monitoring external horizontal displacements at face (topographic marks were located between the blocks 5 and 6 and between the blocks 13 and 14).

Figure 5 also indicates that the wall foundation is composed by a piled slab (concrete platform), due to the presence of soft soil beneath it. **Figure 6** shows the position of the inclinometers (I1A and I2), the load cells used for monitor the reinforcement load, and the total stress cells (C1–C5), located in the first layer of reinforcement at 3.6 m depth. Four load cells were positioned along the reinforcement (see **Figure 7**).

Figure 7. Load cells positioned along the reinforcement [21].

A special device was used for monitoring vertical and horizontal forces at the toe of the bock-face. A bipartite metallic block replaced one of the concrete-blocks that compose the facing (**Figure 8**). Six load cells were used inside this metallic block, four for vertical and two for horizontal load measurement.

Additional details of the instruments used for monitor load in the reinforcements (geogrid) and at the block-face could be found in [21].

2.3. Monitoring results

2.3.1. Tension on reinforcements

Figure 9 shows measured loads in the reinforcement layers at the end of construction (layers 1, 4, 5, and 6, see **Figure 5**). The maximum load recorded was verified in the reinforcement layer 5, and was equal to 7.1 kN/m. Note that the ultimate strength of the geogrid used at the layer 5 was equal to 35 kN/m (**Table 3**). At this layer, the point of maximum tensile load (T_{max}) in the reinforcement at this layer was located 1 m far from face. Notice that considering all layers, the position of the T_{max} does not exhibit a well-defined pattern with respect to the distance from face. This random behavior may be related to the difference of placement of the geogrid and the backfill compaction layers in the field.

2.3.2. Loads at the toe of the wall facing

In **Figure 10**, are shown vertical and horizontal loads measured in the instrumented block located at the toe of the block-face during wall construction. The instrumented metallic block is located in the third block-layer and is monitored by six load cells (see **Figures 5** and **8**).

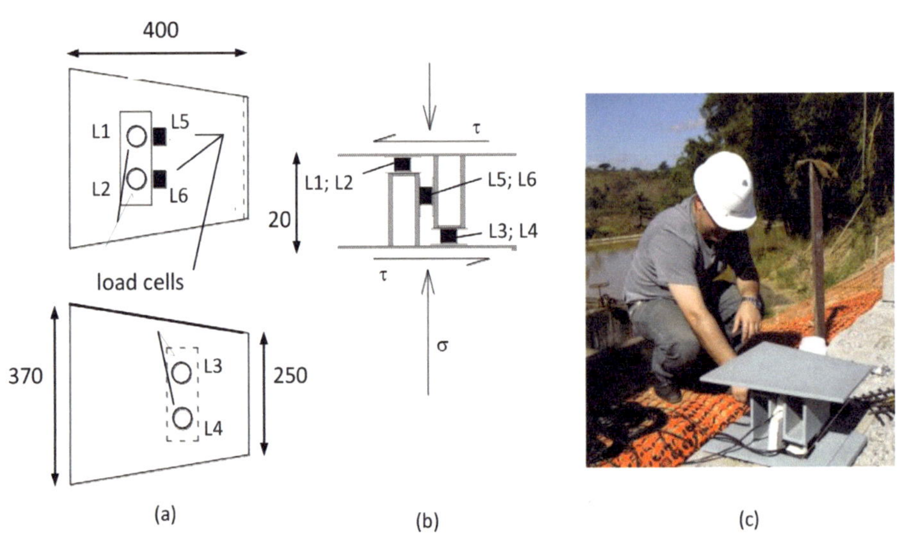

Figure 8. The metallic block used to measure load next to the toe of block-facing: (a) plan view, (b) section view, and (c) block positioned in the field; dimensions in millimeters [6].

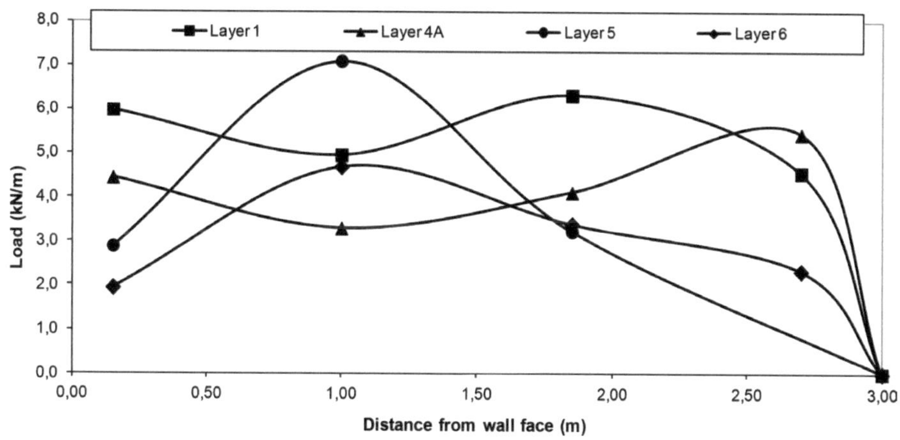

Figure 9. Load in reinforcements measured at the end of construction [6].

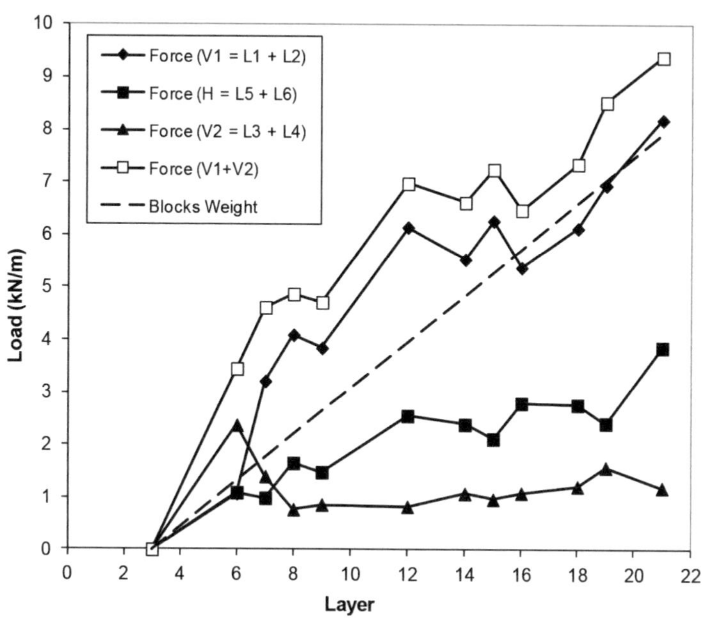

Figure 10. Vertical and horizontal loads measured in the instrumented block during the wall construction [6].

The front (L1 and L2) and rear (L3 and L4) load cells measure the vertical loads acting in the front (V1) and rear (V2) of the block. The load cells (L5 and L6) measured the horizontal load (H) acting in the block. Note that, in **Figure 10**, the front vertical load (V1 = L1 + L2) is often higher that the rear vertical load (V2 = L3 + L4). This behavior is related to the eccentricity of the resultant load due to the self-weight and lateral earth pressure at the interface with the reinforced soil mass that led to an overturn tendency at the block-facing. The dashed line represents the self-weight of the blocks filled with crushed stone, assuming vertical arrangement of the blocks. Notice that the total measured vertical load (V1 + V2) was always higher than the self-weight of the blocks; this increase of vertical load is due to the mobilized friction at the interface of the block-face and backfill. The measured horizontal load at the toe block-face (H) is related to the restrain to the lateral movement at base of the blocks (fix-base condition), as discussed in [22]. Note that in the RSW under analysis, the first block-layer is tied to the concrete slab (see **Figure 2**). At free-base condition, no mobilization of horizontal load at the block-facing would be expected [22–24].

2.3.3. *Vertical stresses at the bottom of the wall*

Figure 11 presents the vertical stress measured by total stress cells (C2–C5, see **Figure 6**) and calculated values using the Meyerhof approach [25] for the first layer of reinforcement (3.6 m depth) at the end of construction. The Meyerhof approach [25] accounts for the eccentricity of the resultant due to the self-weight and the earth pressure exerted by the nonreinforced zone in the wall. The vertical stress provided by Meyerhof [25] is slightly higher than the vertical stress due the self-weight of backfill without any external load. This behavior is due the earth pressure caused by soil behind the reinforced zone. The study carried out by Riccio et al. [6] presents a more deep discussion about this behavior.

2.3.4. *Horizontal displacements*

Figure 12 shows the horizontal displacements measured at the end of wall construction by means of inclinometers (I1A, I1B, and I2; **Figure 5**) and by topographic readings at the end of construction. Significant movements were measured in I1A e I1B near to the face (~60 mm). Topographic readings in the facing at heights of 1.60 and 2.60 m unveil lateral displacements equal to 4 and 22 mm, respectively. The ratio of the lateral displacement in the face and the height of the wall was equal to 1.5%. Moreover, the lateral displacements measured in I2 (nonreinforced zone) were negligible (<2 mm).

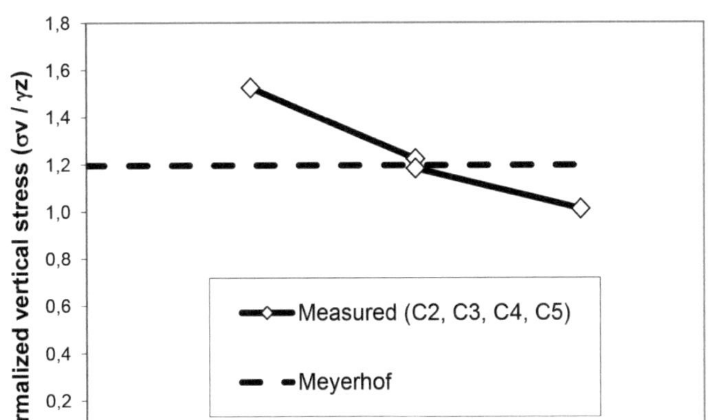

Figure 11. Measured and calculated vertical stress at the base of the wall at the end of construction (third layer).

http://dx.doi.org/10.5772/intechopen.79239

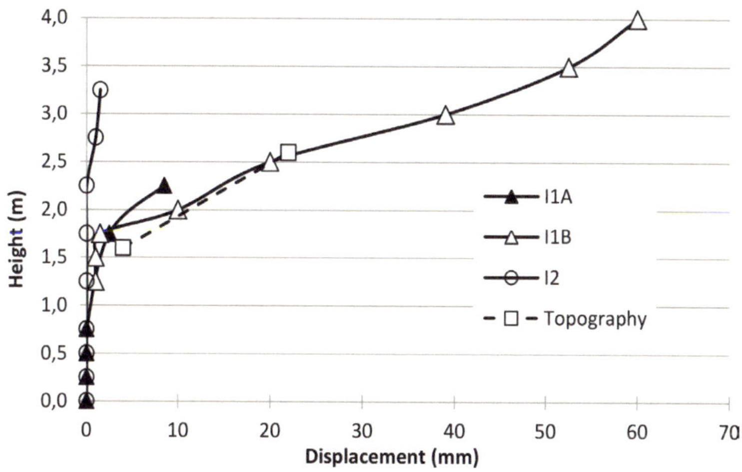

Figure 12. Lateral displacements measured by inclinometers and topographic readings at the end of construction.

2.3.5. *Vertical displacements*

Vertical displacements were measured during and at the end of construction using magnetic settlement plates (P1–P10; see **Figure 5**). Those plates were positioned both in the reinforced zone and the nonreinforced zone. **Figure 13** presents the vertical displacements at the end of construction; the maximum vertical displacement was equal to 18 mm, recorded by the settlement plate P6. Some plates record values equal to zero or less than 2 mm (P4, P7, P8, and P10). Due to the heavy backfill compaction, most of the vertical displacements have occurred during the wall construction. The heavy compaction induces a kind of a preloading of the soil, and it becomes stiffer, preventing additional vertical deformations during the wall service life [11].

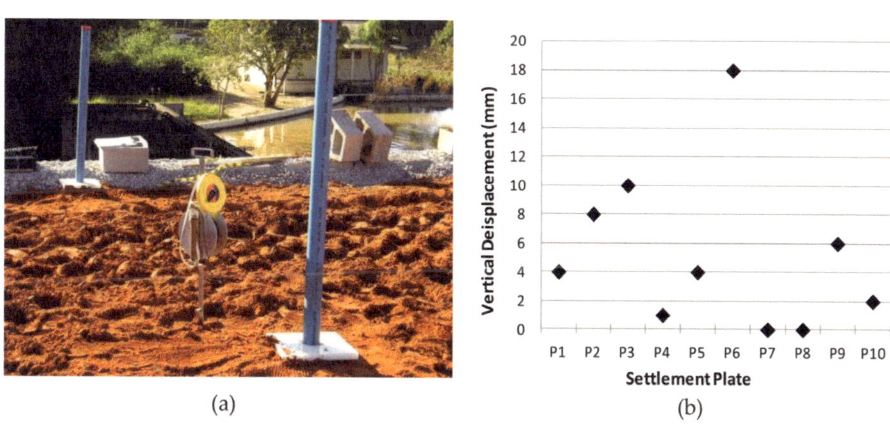

Figure 13. Magnetic settlement plates: (a) view in the field; (b) results at the end of construction.

3. Comparison of measurements and prediction of tension in reinforcements

The basic concept of internal design includes analysis of failure of reinforcement, i.e., it is to verify if the maximum calculated load in the reinforcement (T_{max}) using appropriated method is lower than the design load of the selected reinforcement (T_d). In addition, verification against pullout failure must be done. The design should provide enough length of the reinforcement in the resistance zone (beyond the potential failure surface) to avoid pullout failure. The design strength T_d is estimated at the end of a given reference time (service life) for a particular installation environment and damage that may occur during installation. T_d can be determined by Eq. (1). In this equation, the terms f_f, f_d, and f_a are reduction factors that are dependent of the type of fabric, the service life, the particular installation environment, and damage that may occur during installation.

$$T_d = \frac{T_{ult}}{f_f \cdot f_d \cdot f_a} \tag{1}$$

where

T_{ult} = ultimate tensile strength, i.e., tensile resistance in short-term resistance obtained from the wide-width tensile strength test (the nominal resistance of the geosynthetic);

f_f = creep reduction factor;

f_d = mechanical damage reduction factor;

f_a = reduction factor for chemical and environmental damages.

Table 5 shows the values of T_d and the reduction factors for the installation conditions and geogrids used in the presented wall (see **Table 3**). The reduction values were evaluated considering that: PET geogrid was used as reinforcement; the design service life is 120 years; the pH of residual lateritic soils is around 5 (installation environment); and low damage during geogrid installation (0.30-m thick backfill layers of fine-grained soil and roller drum Dynapac CA250PD). Moreover, in all reinforcement layers, the values of T_d must be higher than T_{max} considering an appropriated factor of safety (FS ≥ 1.5).

Figure 14 shows comparison of measured and calculated load in reinforcements. The determination of maximum load in the reinforcement layers was done using the analytical method presented by Ehrlich and Mitchell [4]. Through this method, backfill shear resistance,

Geogrid	T_{ult} (kN/m)	f_f	f_d	f_a	T_d (kN/m)
1–3	55	1.67	1.05	1.1	28.5
4–7	35	1.67	1.05	1.1	18.1

Table 5. Reduction factors, T_{ult} and T_d values for the fabrics (geogrids) used in the design of the wall.

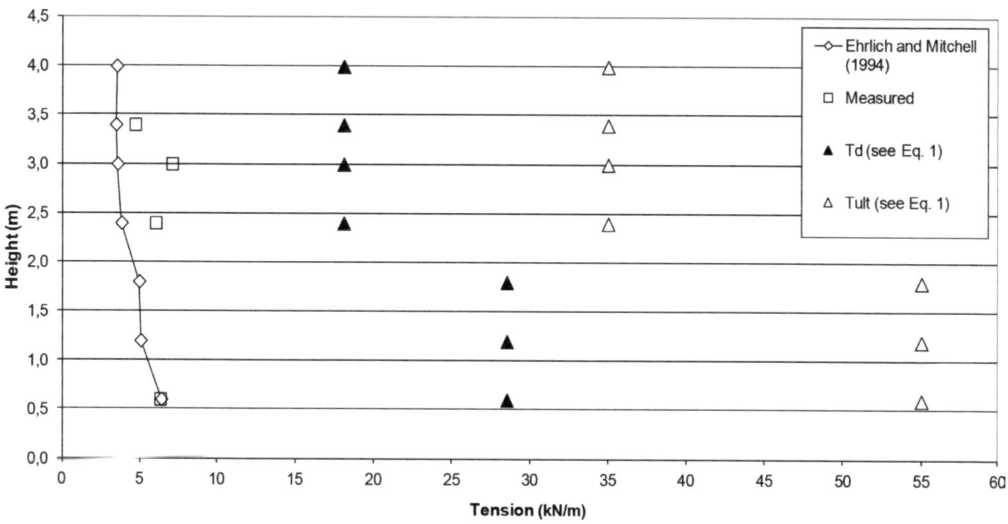

Figure 14. Comparison of T_d and T_{max} measurements and predictions.

reinforcement, and soil stiffness properties are considered, and the backfill compaction stresses are taken explicitly into account. The induced stress due to compaction has the effect of increasing the tension in the reinforcements and the soil cohesion reduced it. In the calculation, the nonconsideration of those factors may lead to poor prediction capability of the real behavior found in the field.

Figure 14 presents that measurements and calculated values of maximum load in the reinforcement (T_{max}) are smaller than T_d. These results also indicate that the predicted values are close to the measured ones, attesting the good performance of the method that was used in the analysis. Additional discussion about measurements and prediction, including determined results using other methods found in the literature, is present in [6].

4. Conclusions

The mechanical behavior of reinforced soil wall built with fabrics (geogrids) is presented based on results of a well-instrumented wall. In this concrete-block-face reinforced wall, tropical fine-grained soils were used as backfill, and two type of fabrics were used as reinforcement. This wall was constructed in 2006 and presents good performance without any structural problem or excessive deformation until nowadays.

Measurements and calculated values of tensions in the reinforcements using an analytical method [4] were compared. Good prediction capability of the used method was verified. In accordance to the good performance of the wall, measurements indicate low vertical and lateral movements, and the mobilized load in the reinforcements was lower than the design load. Measurements also indicate that the block-face supported part of the load that would be carried by the reinforcements.

The fabrics used in the construction were capable to resist the efforts imposed by the structure. The measured mobilized tensions on fabrics (T_{max}) were lower than the design strength (T_d). Considering that T_d is the maximum tension that can act on fabric (T_d is a portion of T_{ult}), it is observed that the wall has safety in terms of internal stability.

Acknowledgements

The authors would like to thank the Brazilian Research Council (CNPq) and HUESKER Synthetics for the support for this work.

Author details

Mario Riccio[1]* and Mauricio Ehrlich[2]

*Address all correspondence to: mvrf1000@gmail.com

1 PEC-UFJF, Federal University of Juiz de Fora, Juiz de Fora, Brazil

2 COPPE-UFRJ, Federal University of Rio de Janeiro, Brazil

References

[1] AASHTO. AASHTO LRFD Bridge Design Specifications. 7th ed. Washington DC; 2014. p. 2150. ISBN: 9781560515920

[2] BSI. BS 8006-1: Code of Practice for Strengthened=Reinforced Soils and Other Fills. London, UK: BSI; 2010

[3] Bathurst RJ, Miyata Y, Nernheim A, Allen AM. Refinement of K-stiffness method for geosynthetic-reinforced soil walls. Geosynthetics International. 2008;**15**(4):269-295. DOI: 10.1680/gein.2008.15.4.269

[4] Ehrlich M, Mitchell JK. Working stress design method for reinforced soil walls. Journal of Geotechnical Engineering, ASCE. 1994;**120**(4):625-645. DOI: 10.1061/(ASCE)0733-9410(1994)120:4(625)

[5] Ehrlich M, Mirmoradi H. A simplified working stress design method for reinforced soil walls. Géotechnique. 2016;**66**(2):854-863. DOI: 10.1680/jgeot.16.P.010

[6] Riccio M, Ehrlich M, Dias D. Field monitoring and analyses of the response of a block-faced geogrid wall using fine-grained tropical soils. Geotextiles and Geomembranes. 2014;**42**:363-374. DOI: 10.1016/j.geotexmem.2014.01.006

[7] Stuedlein W, Allen T, Holtz R, Christopher B. Assessment of reinforcement strains in very tall mechanically stabilized earth walls. Journal of Geotechnical and Geoenvironmental Engineering, ASCE. 2012;**138**(3):345-456. DOI: 101061/(ASCE)GT.1943-5606.0000586

[8] Mimorardi H, Ehrlich M. Numerical evaluation of the behavior of GRS walls with segmental block facing under working stress conditions. Journal of Geotechnical and Geoenvironmental Engineering. 2015;**141**(3):04014109. DOI: 101061/(ASCE)GT.1943-5606.0001235

[9] Mimorardi H, Ehrlich M. Investigation of the prediction capability of the EM design method under working stress conditions. In: Proceedings of GeoShanghai 2018 International Conference: Geoenvironment and Geohazard. GSIC 2018; Springer, Singapore. pp. 527-536

[10] Ehrlich M, Mimoradi H, Saramago R. Evaluation of the effect of compaction on the behavior of geosynthetic-reinforced soil walls. Geotextiles and Geomembranes. 2012;**34**: 108-115. DOI: 101016/j.geotexmem.2012.05.005

[11] Ehrlich M, Mitchell JK. Working stress design method for reinforced soil walls- clousure. Journal of Geotechnical Engineering, ASCE. 1995;**121**(11):818-821

[12] Mirmoradi SH, Ehrlich M. Modeling of the compaction-induced stress on reinforced soil walls. Geotextiles and Geomembranes. 2015;**43**(1):82-88. DOI: 10.1016/j.geotexmem.2014.11.001

[13] Carvalho PA, Wolle CM, Pedrosa JABA. Reinforced soil embankment using geotextile, an alternative for geotechnical engineering. In: Proceedings from 8th Brazilian Conference on Soil Mechanics and Foundation Engineering; Porto Alegre, Brazil; 1986. pp. 168-178, (in Portuguese)

[14] Ehrlich M, Vianna AJD, Fusaro F. Performance behavior of a geotextile reinforced soil wall. In: Proceedings from 10th Brazilian Conference on Soil Mechanics and Foundation Engineering; Foz do Iguaçu, Brazil; 1994. p. 819e824 (in Portuguese)

[15] Bruno AC, Ehrlich M. Performance of a geotextile reinforced soil wall. In: Proceedings from 2nd Pan-American Symposium on Landslides; Rio de Janeiro, Brazil; 1997. pp. 665-670

[16] Ehrlich M, Delma V, Carvalho P. Performance of two reinforced soil slopes. In: Recent Developments in Soil and Pavement Mechanics. Balkema; 1997. pp. 415-420

[17] Benjamin CVS, Bueno BS, Zornberg JG. Field monitoring evaluation of geotextile-reinforced soil-retaining walls. Geosynthetics International. 2007;**14**(2):100-118. DOI: 10.1680/gein.2007.14.2.100

[18] Brugger PJ, Schmidt CF, Hemsi PS. High height GRW with green wrapped facing. In: Proceedings from REGEO/Geossintéticos; Belo Horizonte, Brazil: CD-ROM (in Portuguese); 2011

[19] Duncan JM, Byrne P, Wong KS. Strength, Stress-Strain and Bulk Modulus Parameters for Finite Element Analyses of Stresses and Movements in Soil Masses. California, USA: University of California, Berkeley; 1980; Report UCB/GT/80-01

[20] Ehrlich M, Becker LB. Reinforced Soil Walls and Slopes, Design and Construction. 1st ed. Boca Raton: CRC Press; 2009. p. 126

[21] Riccio MVF, Ehrlich M. Monitoring of a geogrid reinforced soil wall with segmental block facing. In: Proceedings from Proc. GeoAmericas; Lima, Peru.CD-ROM; 2012

[22] Mirmoradi SH, Ehrlich M. Evaluation of the combined effect of toe resistance and face inclination on the behavior of GRS walls. Geotextiles and Geomembranes. 2016;**44**(3):287-294. DOI: 10.1016/j.geotexmem.2015.12.003

[23] Leshchinsky D, Vahedifard F. Impact of toe resistance in reinforced masonry block walls: Design dilemma. Journal of Geotechnical and Geoenvironmental Engineering, ASCE. 2012;**137**(2):236-240. DOI: 10.1061/(ASCE)GT.1943-5606.0000579

[24] Ehrlich M, Mirmoradi SH. Evaluation of the effects of facing stiffness and toe resistance on the behavior of GRS walls. Geotextiles and Geomembranes. 2013;**40**:28-36. DOI: 10.1016/jgeotexmem.2013.07.012

[25] Meyerhof GG. The bearing capacity of foundations under eccentric and inclined loads. In: Proceedings from 3rd International Conference on Soil Mechanics and Foundation Engineering; Zurich, Switzerland; 1955. pp. 440-445